A Treasury of Great American Houses

by Henry Lionel Williams
and Ottalie K. Williams

G. P. Putnam's Sons New York

First Edition 1970
All rights reserved. This book, or parts thereof, must not be reproduced
in any form without permission.
Published on the same day in the Dominion of Canada by Longmans
Canada Limited, Toronto.
Library of Congress Catalog Card Number: 70-116993.
Printed in Switzerland.

Contents

Contents

Preface

This volume presents a group of thirty of the finest mansions erected in the continental United States between the mid-eighteenth and the early twentieth centuries. Culled from twenty states, these represent houses of a wide variety of architectural styles from Early Georgian to Late Victorian, plus late nineteenth- and early twentieth-century versions of designs current in England, France, and Italy at much earlier periods. A notable omission, however, is the absence of any house that can properly be characterized as historic in the sense that it has been directly associated with either notable events or personages in the country's past.

On the other hand those included here comprise a countrywide selection of truly great houses, the finest if not always the largest that this country has produced, and possessing interiors to match the quality, if not the period, of their architecture. In most instances the authors were able to inspect these mansions at first hand in the course of a 17 000-mile journey made for that purpose, as a result of which it has been possible to include a gratifying proportion of important private residences, in addition to others owned by organizations devoted to their preservation.

In several instances access to these great houses was made possible only by the kind cooperation of certain individuals who not only knew of the existence and whereabouts of properties worthy of inclusion, but were happy to effect the necessary introductions to the owners thereof without any prior commitment on the part of the authors whose sincere thanks, accordingly, are due to their good friends: Mrs. Katharine B. Dehler, Mrs. Fred L. Ingate, Mr. and Mrs. John M. Kohler, Mr. John C. McGuire, Mrs. Walter Scott, Mr. Bruce Power, Mrs. Enid T. Thompson, and Mr. Jerome Zerbe.

H. L. W. and O. K. W.

Introduction

In discussing the following group of great houses, it is necessary to emphasize the fact that, for present purposes, greatness is neither equated with size nor considered synonymous with elaborateness of design, splendor of interiors, or even the magnificence of the surrounding landscape. Rather the distinction depends upon the degree to which the house reflects the owner's social status while meeting his practical needs and satisfying his aesthetic tastes. And it is remarkable how consistently such requirements can be fulfilled even in the cases of mansions that have served as cherished homes for a period of several generations, although the structure may have been modified and the ownership fallen into other hands.

In the following pages, then, a serious attempt has been made to select examples of varied architectural styles, regardless of size, covering both original period houses and later copies or adaptations—not omitting two examples of unique design—ranging in date of erection from the early eighteenth century to the first quarter of the twentieth. In certain instances the architects or the owners themselves, having toured much of Europe and the British Isles, selected favorite prototypes which they could modify to suit their taste and convenience, as well as their pocketbooks, with sometimes astonishing results. The houses themselves range from comfortable, though capacious homes to palatial residences that seem designed solely to impress, some of them having remarkable histories, while others constitute nostalgic reminders of both the distant and more recent past, inviting comparisons that are fascinating in the extreme.

Up to the early part of the twentieth century, regional characteristics seem to have played little part in the design of the larger mansions, either with regard to the types of materials used or the styles of architecture. Each owner or architect seemingly preferred to copy or adapt some residence that struck his fancy either here or in some foreign country, apparently leaning toward Renaissance styles and their national

Cupola House, Edenton, North Carolina. Georgian cupola added 1758.

equivalents. Quite often they built these houses of materials foreign to the area—and sometimes to America—so that the design has little relationship to either time or place. Fortunately, size is often combined with a certain modesty of character, and the best houses, as distinguished from the most flamboyant, likewise may possess a richness of style that never degenerates into vulgarity, mere ornamentation escaping confusion with interesting design.

Of equal importance in determining the interest of any specific dwelling, from the reader's viewpoint, is the possible discrepancy between the exterior architecture and the interiors, one or the other being quite unrepresentative of its class—which is not to say that every house included here is a monument to architectural perfection either inside or out, according to any century's standards. In fact there are many mansions in which the interiors are of a far different period or style than the exteriors, a trend that is an offshoot of the eclectic style carried to its logical conclusion.

Even in these pages where the accent is on the best available house of its kind, one may discover that an architect can pay too high a price for a perfectly balanced façade, achieved without regard to varying room sizes which dictate the location of windows and doors. In like manner, the designer can often encounter difficulties in trying to combine the attractive features of two or more styles of mansions, or incorporating more recent features in a house of an earlier date. A good example of this would be the well-known Cupola House in North Carolina (not dealt with here), which combines a New England Colonial overhang with external end chimneys of Virginia origin and a fantastically oversized cupola behind a heavy cross gable or pediment, both of which favor the Georgian. A century later, when the popularity of the Greek Revival style was on the wane, owners were still elaborating plain two-story houses with classical porticoes that sometimes overpowered the rest of the structure.

On the other hand there are many utterly appealing houses that it is quite impossible to type, such as the Schatz house at Helena, Arkansas, which was among those most regrettably omitted. This is a lovely mansion, full of charm, but it was inescapably difficult to fit it into any logical category, full as it was of fine furniture, antique and modern, from almost every country in the world. This house is much larger than it appears from the enormous front lawn, incorporating details from both Roman and Late Palladian styles—with probably the only example extant of a massive Palladian window mounted on a ridge roof, cheerfully exploiting an entirely unbalanced façade whose flat porch roof is carried around one end and a fully exposed basement under an apparently endless panel-and-baluster-railed terrace that continues beyond the house, disappearing under trees in the distance.

Other examples show how domestic architecture can vary both in style and interpretation whether the changes are made

Estevan Hall, Helena, Arkansas.
Unique design, 1810–1820.

for change's sake or provoked by problems forced upon the designer by practical considerations. Two extremely interesting examples illustrate the possibilities innate in the adaptation of a sixteenth-century English-style house to differing purposes, one palatial in concept, the other a strictly country residence of unusual plan laid out to accommodate museum-quality art and artifacts acquired during a lifetime of collecting.

In one house of fascinating interiors, a Greek Revival front revolutionizes a Spanish Colonial dwelling to the benefit of both. In another, Mediterranean, Palladian, and Southern Colonial styles are melted into an exquisite whole, while a third successfully combines English and French characteristics, adapted to the Palladian five-part plan, with Adam touches, to provide something entirely different in appearance and accommodation. Probably the most unusual of all, however, is an owner-designed, stone-faced mansion developed from the inside out and employing, for the first time in recorded history, hollow walls of reinforced concrete for a domestic edifice that has proven both utterly practical and the acme of comfort and workability, sitting on its rugged mountainside as though it had been there since time immemorial.

These and the rest of the mansions described in this book represent almost every period and every style—or acceptable combination of styles—to achieve the approval of architects and the general public alike, which is not to say that there is always a clear relationship between a mansion's architecture and its furnished interiors. Good taste in either is far from universal, and allowances have to be made for a variety of factors, practical or sentimental, that may influence decisions. Not everyone is willing to discard, or even hide, an ancestor's treasured possessions.

One other point to be emphasized is that each architectural style is subject to different interpretations in different countries, so that all English Georgian houses and Italian Renaissance *palazzos*, for example, built in America will normally display some American quirk in regard to details, for the same reason that Chippendale furniture made in Philadelphia or New England is easily identifiable as such. Similarly, detail differences will be found in houses of the same style erected some years apart, thanks largely to fluctuations in fashion, variations in the competence of the architects involved, the desire for individuality, and the necessity for compromise because of rising costs or changes in specifications, while the builder who is his own architect may sacrifice external grace for impressive interiors—or vice versa.

In this connection it is remarkable how the medieval, of whatever country, appeals to some American home owners— or at least the more romantic among them. It is also fortunate that the architects are, for the most part, able to translate such divagations into a modern idiom, producing something that pleases when adapted to today's circumstances and applications, without lowering standards to a point where they

T. Macdonough Russell House,
Middletown, Connecticut.
1850 Gothic Revival.

Woodlawn Plantation,
Mount Vernon, Virginia.
Washington's wedding gift to
Nelly Parke Custis.

are quite out of keeping with modern equipment and facilities, or introducing marked crudities and their accompanying discomforts, whether physical or psychological.

Turning to the matter of interiors, any room's architectural details which provide the bare bones of a background for the furniture may range from the elaborate to the meager. On the other hand it may be the more or less permanent decoration of the walls and ceiling in the manner of Adam or Wyatt or the earlier Italian Renaissance period. These would be the fresco designs (or possibly oil paintings) that determine the room's character, and to which the furniture style must either conform or contrast, depending on the effect desired. In some instances it is even possible to make up in decoration (or interior architecture) what is lost through the reduction of furniture to mere essentials, thereby achieving a lively and attractive interior.

In any event it is interesting to compare the obvious objectives of the present-day owners in rooms recreated from old-time furniture listings with those of more recently constructed mansions in which the owners have followed their own inclinations, more often than not adding inherited pieces, modern seating, and other preferences in order to achieve both comfort and convenience according to modern standards.

Such an atmosphere will usually be governed by the tendency of the owners (former or present) to duplicate the practice of some specific period. On the other hand it often happens, especially in houses open to the public, that the rooms cater to the modern idea of what any such interior should be like. Thus it may happen that in the antique houses pictures are not hung as high as they should be for authenticity's sake, while carpets and rugs are placed on the floor instead of on tables or hung on walls. Likewise, in the Late Victorian era rooms are not as properly crowded with furniture (not to mention potted palms!) as they should be. Instead they are scantily furnished as a sort of psychological protest against what is taken to be the obviously poor taste of an era now in disrepute.

In spite of all such possible gaffes and gaucheries, the rooms shown here do constitute a fairly accurate pictorial essay on at least one aspect of the way of life of well-to-do persons of widely dissimilar backgrounds and cultures, whether native or foreign-born, who, with all the world to borrow ideas from, have, with or without the help of an interior designer, given expression to their own tastes and predilections for various times, long past or more recent, to which their homes are today a living witness.

In the following pages, then, are shown and discussed some of the better known and some far less familiar examples of architectural styles and furnished interiors over a period of 200 years ending about four and a half decades ago. These have been selected from widely spaced areas throughout the country, each constituting a contribution to our cultural history that can never be replaced. For this reason the nation's gratitude is not only due to their present individual owners, but also to those organizations who have made it their business to see that these great houses and their priceless contents are preserved for posterity as a cultural asset that can never be duplicated.

Mayfield, Chester County, Pennsylvania. An eighteenth-century Greek Revival mansion.

Wilton (built 1750) after its removal to Richmond.

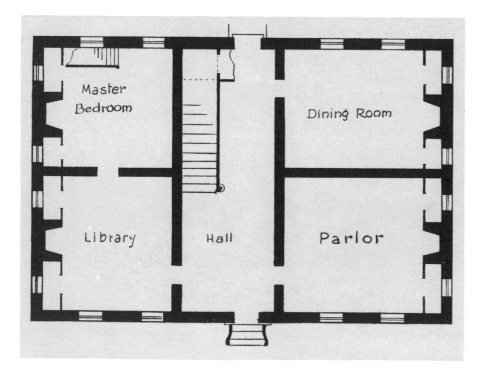

The main floor of Wilton.

WILTON

From World's End to Windsor Farms

Built in 1750, Wilton is a remarkable house with an astonishing history—remarkable because it is paneled from floor to ceiling in every room, hall, and closet, and floored in heart pine; astonishing because 180 years after its completion at World's End, overlooking the James River at Falling Creek, it was rebuilt atop another James River bluff at Windsor Farms, Richmond, seven miles from its original site.

Although no one knows who designed the house, it is believed that the owner, William Randolph III, was both the architect and builder, perhaps basing his plan on Westover, the earlier Byrd mansion, or adapting it from Salmon's book *Palladio Londinensis*, though some think that the architect Richard Taliaferro (1705–1779) may have had something to do with it. In any event it is a dignified building of masculine simplicity in the Early Georgian style, nice in proportion and possessing some of the loveliest architectural interiors ever designed—especially of this period when all important rooms were painted.

Wilton originally had five dependencies, including the all-important kitchen. Today, on its new site, and restored to its original condition by the National Society of the Colonial Dames of America in the Commonwealth of Virginia, it has none. The house itself has nine windows to the two-story principal façade and a low-pitched hipped roof with two pairs of massive, widely spaced end chimneys flush with the gables and tall enough to rise well above the ridge. The bricks are laid in Flemish bond, the window frames set within the reveals but given added importance by rubbed-brick areas—a brick and a half wide—down each side and across the flat arches, formed of voussoirs that span the tops. The pilastered and pedimented doorway likewise is emphasized, as are the heavy stringcourse and the building corners, where the rubbed areas appear, from a distance, to be adorned with quoins.

The whole structure is set up sufficiently high to expose about eighteen inches of the basement windows, with half a dozen iron-railed stone steps leading up to the rather sedate entrance—

A corner of the master bedroom showing nursery stair.

The entrance hall with portraits of Elizabeth and Beverly Randolph.

its pediment modillioned like the eaves cornice—with sill-height plantings of boxwood hiding the foundations and water table, thereby adding to the illusion of length. Its location on a high bank, at the top of a flight of brick steps, helps endow the mansion with an imposing air worthy of its beautifully landscaped setting.

Considering Wilton's unpretentious exterior, a view of the interior through the open doorway is little short of astonishing. Here is a 14-foot-wide hall, 27 feet long to the foot of the stairs, alongside which it runs back another 12 feet to the garden door overlooking the river. This door, like the front one, is close to the right-hand wall—a consequence of the hall being off-center so that the rooms to one side of it could be made larger than those at the other side without increasing the overall length of the house. Unfortunately, no window could be placed alongside either door to help illuminate the hall, since this would unbalance the façade and spoil its Georgian character. In spite of this, the first impression to be gained on entering is of a beautifully wainscoted room, warm and welcoming in a lovely shade of green.

The paneling, accented by a heavily molded dado rail, sweeps up the stairs which, by means of two crossover landings, turn back upon themselves above the rear-door transom light, exposing a paneled soffit as they reach up to the floor above. This stair and the balustrade are the only sections of woodwork

16

in this interior left with a natural-wood finish. Near the garden door is a Windsor-style settee which invited the exchange of riding boots for indoor footwear from the closet formed by enclosing the stairs leading to the basement. Farther along, a Chippendale side table beneath a looking glass faces, across the hall, a splendid eighteenth-century tall clock by Simon Willard, flanked by portraits of Elizabeth and Beverly Randolph, while spoon-footed Queen Anne chairs are placed around for convenience.

In the corner of the first stair landing is a handkerchief table with a brass candlestick, six more of which are arranged on ledges in the paneling above. The low slat-topped door to the basement stairs faces a normal-sized entrance to the dining room, the two of them used in the early days for service from the outdoor kitchen. Of two other doors at the front end of the main hall, one gives access to the parlor, the other to the library from which a doorway in the transverse wall opens into the master bedroom.

A peculiar feature of all these rooms is the masking of the recesses on either side of the fireplace in which the end windows are located. In the case of the parlor, the chimney breast paneling is flanked by two arched openings, each revealing a narrow window, but in the dining room, similar recesses, ordinarily concealed by doors, allow the deep window bottoms to be used as serving counters.

Deep window recesses flank
the parlor fireplace.

The Washington Room with
Peale's portrait of the general
in his earlier days.

At the opposite side of the hall, the spaces alongside the chimney breasts serve, in the case of the library, as book storage, and those of the bedroom as clothes closets, though that to the right of the fireplace loses considerable space to a narrow stairway which leads to the nursery on the upper floor. This flight of stairs, incidentally, has all the earmarks of a later addition, cutting, as it does, across the lower part of one of the rear windows, but the base of the newel-post found in the original flooring is taken as proof of its being part of the original structure.

As might be expected, the parlor is undoubtedly the mansion's finest interior, both from an architectural and a decorating standpoint. The ceiling, being flat and white, serves to accentuate the four pilaster caps at each end of the room, which are topped by full entablatures, the cornices projecting, *en ressaut*, at ceiling level, where they are duplicated by similar upward extensions of the arch keystones. The archways, too, are quite elaborate, set between fluted pilasters with paneled spandrels and molded imposts. The white-marble fireplace is beautifully carved in basso-relievo, with side panels and a heavy keystone held within ogee curves, all on a white-marble hearth.

In the arched recesses, single draperies of scarlet damask are looped to one side by heavy, tasseled cords, making room for inlaid mahogany gaming tables, each displaying a piece of rare porcelain and a silver candlestick. Around the grouped Oriental rugs are finely carved Philadelphia Chippendale chairs in red silk damask; a wing chair in yellow silk, with vertically rolled arms; a beautifully carved, piecrust-top, birdcage type of tea table; a small octagonal toddy table; and similar eighteenth-century pieces, on all of which an overmantel portrait of William Randolph III, attributed to John Wollaston, smiles benevolently down.

In the dining room the green-paneled woodwork is similar, though the arches are replaced by rectangular panels over flat-topped doors. The fireplace, however, is a duplicate of that in the parlor. Above it is a portrait of Mrs. Peyton Randolph, and to one side of the unusually long marble hearth is a three-tiered dumbwaiter laden with a silver tankard and china and glassware of the period. The Oriental rug here is large enough to accommodate the round-ended Hepplewhite dining table, fully extended, and a set of Queen Anne vase-backed chairs with bell-type seats, while against the nearby wall is a nice inlaid satinwood Hepplewhite sideboard over which hangs a Chippendale-style looking glass in gold leaf.

Across the hall, both the library and the master bedroom have quite plain, flush-style, white-marble fireplace mantels, that of the library flanked by fluted pilasters plus entablatures, the bedroom paneling remaining quite plain. In the library a portrait of William Randolph II hangs over the mantel, in front of which is arranged a quite masculine furniture grouping consisting of a folding gaming table with a Queen Anne base-back chair on one side and a Hepplewhite shield-back on the other in green and gold medallion-patterned brocade. On the opposite end of the small but elegant Oriental rug, a round-topped tripod table serves as candlestand and bookrest, alongside a wing chair in brass-nailed damask of tan and gold in a large floral pattern. The windows have deep pelmets and fringed draperies of gold damask, the same material being used to cover the cushions in the recessed window bottoms.

The master bedroom next door is quite gay in contrasting red seat coverings, deep blue on white-figured, resist-print draperies, bed tester, and hangings in an oversize pattern, and an embroidered heirloom bedspread in white, against the pale-green paneling and the wood tones of furniture and floor, with which the brownish-pink grounds of the Oriental rugs have so much in common.

This is a small but charming room with an elegant spoon-footed corner chair with pierced splats; a straight-legged Chippendale side chair used with a slant-top desk; a nicely carved, dish-top tripod table; a folded Hepplewhite card table, and a canopied bed with stop-fluted tapered posts, plus a small fireplace hearth occupied by a brass teakettle on a Queen Anne footman. Other gold tones are provided by gilt-framed prints, polished candlesticks, and the desk writing set which sports a pair of goose-quill pens.

In the spacious upper hall, with its ceiling-high paneling

The dining room with its portrait of Mrs. Peyton Randolph.

In the library William Randolph II looks down on a gaming table set for play.

A corner of the smallest bedroom.

painted the same green as the lower hall, both the dado rail and the cornice are reduced to a minimum. At the front end, where there is no stairway to interfere, the single, offset window leaves sufficient wall space for a few furniture pieces, such as a slant-top desk and its Chippendale chair, even though the doorways to the side rooms are a mere five feet from the front wall.

In this area the furniture grouping is tied together by an Oriental rug occupied by one of a pair of Queen Anne gaming tables with sinkings for candlesticks. This is placed under the window and accompanied by an American cane-back armchair and a side chair, both dating from about 1700. This window is draped, almost flush with the reveal, in a fringed print that forms a double swag at the top and sill-length cascades at the sides. Two other Oriental rugs break up the floor area, the longer one fronting a high-backed Sheraton sofa of 1790 to 1800 with reeded legs, the smaller rug serving a huge, eighteenth-century, English flat-topped mahogany bookcase. While this bookcase is flanked by straight-legged, eighteenth-century mahogany side chairs, the sofa is limited to a Chippendale chair at one end and a folded duplicate of the baize-topped gaming table at the other.

Most of the upstairs bedrooms are quite simple both as to background and furnishing, having the same kind of simple paneling in a lighter tone of green than that of the main-floor rooms. The flush fireplace surrounds, however, are of black marble, with a lightly carved rosette at the center of the frieze or lintel; there is no shelf. The most completely furnished upstairs bedroom is that known as the Washington Room, the general having visited Wilton several times, though whether he occupied this room or not is a matter of conjecture. In any event the overmantel is graced by a copy of Charles Willson Peale's portrait of the great man in his younger days, wearing a gorget and the uniform of the Virginia militia.

Today the right-hand fireplace recess is furnished with a corner washstand, fully equipped with basin, pitcher, and hot-water jug. On the other hand the tripod basin stand is in the bedroom proper, along with a brass warming pan. There is also another of the slant-top desks—an original Randolph piece—together with its shield-back armchair, against the same wall as the four-poster bed, which has a high, exposed back, suggesting a later period. The canopy, however, is of a tailored pattern in an early print, lined with white cotton fabric, as are the tie-back side curtains. This bed has an open foot and is low enough not to require the use of bed steps. On the floor are the usual Oriental rugs, the hearthside one affording room for a bedroom-sized wing chair, together with a round-topped tripod table large enough to accommodate the usual candlestick, Bible, and spectacles.

The smallest of the bedrooms is quite similar in layout and furnishing, though the bedposts are somewhat shorter and terminate in wooden finials. The carved posts are clustered for half their length in the Chippendale Gothic style, with a

ruffled tester valance and dust ruffles reaching the floor, while back draperies hide the bedhead.

Through the open door of the fireplace recess can be seen a raised-back washstand, the candle shelf surmounted by a looking glass. On the stand itself is a basin and pitcher, with a shallow drawer for accessories below the top. Built-in shelves above the washstand are used for storing linen.

In the bedroom is a large hooked rug in a stylized floral design that goes well with the bed draperies and the silk-damask covering of the high-backed, upholstered, fireside "lolling" chair. This latter piece has little in common with the Philadelphia Chippendale bedside chair, though both of them together serve as a reminder of the changes in furniture styles that could take place during a century of continuous occupation by the same family in a day when strictly period-style interiors were far less important than they were to become in the classical period that lay ahead.

In the huge upper hall the offset window leaves space for a writing desk.

THE LINDENS

Doubled in Size, Trebled in Beauty

The Lindens framed by its giant linden trees planted in 1850.

One of the loveliest façades in all New England is that of the house built in 1791–1793 by Julius Deming at Litchfield, which has remained a town of beautiful homes to this day. Unlike most houses of its period, the Deming place seems to have grown even more splendid with the passage of time, thanks to the care with which successive owners adapted it to their needs.

The house remained in the Deming family until 1910; and it is an interesting fact that in 1850 Mr. Deming's daughter, Lucretia, planted the row of linden trees still growing in front of it—from which the property more recently took its name.

As originally designed by William Sprats (1747–1810)—a Scottish master builder who became a prisoner of war during the Revolution—the mansion was a two-and-a-half-story clapboarded structure with a lookout deck between the pair of twin chimneys that graced its hipped roof. Considerable character and not a little charm was supplied by the heavy rusticated quoins and the central pavilion with its pedimented roof over a lovely Palladian window, supported by a neat blind balustrade of equal width and enframed by fluted Ionic pilasters—all above a portico in the same classic style which sheltered a heavy paneled door with a rectangular fanlight and well-spaced sidelights typical of Late Georgian days.

Another attractive feature, not often met with elsewhere, lies in the heavy modillioned and dentiled hoods installed over the narrow side windows of the Palladian unit and the four first-floor windows, those of the inner pair being carried right around the portico in the form of a cornice. All of these projecting moldings are painted white like the rest of the façade, showing to advantage the buff front door and the green shutters.

Insofar as the principal façade is concerned, a major change made in recent years was the substitution of a mansard roof in place of the original hipped one to provide more space on the attic floor. This turned the rooftop into a large, square deck, now balustraded to include the end chimneys, the front slope

An 1890 rear addition based
on the house's original features.

Even in the entrance hall
the architectural features
are notable.

A shell-topped cupboard displaying pewter is a dining room feature.

Heavy moldings give character to the parlor fireplace.

incorporating two hip-roofed formers, straddling the pavilion pediment, with four on each side of the main roof.

The house was enlarged about 1890 by a nephew of Julius Deming's. Since then other additions have been made from time to time, particularly in the matter of accommodations for the staff, but the original construction has been little altered.

These appendages and extensions, which naturally affect the appearance of the residence from every angle, have resulted in a unified architectural composition that is equally enchanting from whichever side it is observed. Undoubtedly the most ambitious project was the erection of a two-story Ionic portico at the rear of the house—an almost exact copy of that of the Cowles house nearby, including an arched foundation wall necessitated by the high raised basement. At the same time a balcony was installed above the rear door, with access from a second-floor French window, adding to the romantic aspect of the whole structure.

From this flagged portico one can either descend a wide, L-shaped, iron-railed stairway to the lawn or move around on the portico to a small side terrace in the angle of the house formed by a narrower addition to the original building. Here a pleasant, sheltered nook is formed by a room-high bow window, decorated with exquisite moldings, at right angles to the rear corner of the old house, where a French window gives access to the living room.

With all of these beautiful details worked so logically into the exterior of the mansion it is no surprise to find the interiors also replete with architectural features that not only enrich the rooms but also endow each one of them with a character of its own, regardless of the type and style of furniture used. Naturally, as befits the house's long history, this includes both antique pieces and reproductions, together with modern pieces so necessary to meet today's standards of comfort.

Much of this effect is due to the fact that the architectural features are bold and positive in the larger rooms and far from neutral in the smaller ones, a situation that presents problems in furnishing. Fortunately, the present owner has been able to balance the contents against the background by the simple expedient of either exaggerating or minimizing the effect of the architectural features by the adroit use of color and variations in the degree of contrast in texture and pattern. All of this can clearly be seen in the contrasts provided by the morning room and the parlor.

In both of these rooms the fireplace is the one outstanding feature with its heavy molding—the huge pillow block in one, the row of large triglyphs in the other, both sharing the impact of the multiplicity of modillions that not only emphasize the cornices of the mantelshelf and pediment, but are carried right around the room under the main cornice. In the morning room all the woodwork—the entire fireplace wall, the paneled dado, doors, and window trim—is painted an antique blue.

Once a kitchen, the dining room
is now the most formal interior.

In the parlor there are the same heavy architectural features, this time with a complete pediment above the overmantel panel and the same deep cornice. But here the fireplace wall retreats a little because of the cream panels in a pale-beige molding.

The impact of the molding is further reduced by the crimson brocade used for the voluminous hangings, whose thick pelmets are formed of a double swag of the same material. Furthermore the overmantel is given a definite focal point by a circular bull's-eye mirror in a gilt frame centered on it, between a pair of antique French urns in green and gold. All these are contrasted with a Phyfe sofa in a deep-blue cotton fabric; an eighteenth-century wing chair in a light French floral pattern on a dark-blue ground; and mahogany furniture with an 1850 Litchfield-made papier-mâché table in black and gold; a tall Sheraton bookcase-desk in banded mahogany; and Chippendale chairs covered in a pale-blue silk with a pattern of white squares in trellis fashion.

One of the loveliest interiors of this remarkable house is that of the dining room, which is the antithesis of all the foregoing. It combines an extraordinary variety of colors with natural woods and is quite a formal room as compared with the others, which tend to emphasize that lived-in look. Much of the wall area is paneled in stained pine, including the raised-panel dado. Those wall areas not covered by paneling are

A beautiful bookcase adapted to the hall by removing its finials.

A wide Palladian window sets off this cozy retreat in the upper hall.

papered in a gay floral creeper design of greens, reds, and yellow, exposing a great deal of the neutral background.

The room itself is somewhat narrow—it was once the kitchen—but luckily the fireplace is toward one end so that there is ample room for the large formal dining table—a twin pilastered style, each with four feet—and the spade-footed eighteenth-century Sheraton sideboard. At the other end of the room, opposite the fireplace, is a shell-topped cupboard displaying pewter pieces, let into a projection in the paneled wall made to receive it. This further reduces the width of floor available for a smaller square table of the same four-legged pedestal type used for family and other informal meals.

This table faces a wide window adorned with a rectangular, tailored pelmet over hangings which, like the room's other curtains, are of twill with a crewel embroidery, the work of three generations of women in the owner's family. At the other end of the room is the kitchen doorway, hidden by a threepanel, crewel-embroidered screen. By its side stand an English eighteenth-century flat-topped vitrine, its shelves loaded with chinaware, and against the other wall is a drop-leaf serving table.

A vastly different interior in every way is represented by the huge living room, where architectural features encompass the four walls, yet seem in not the slightest degree insistent. What they do accomplish with their continuous panel work of dadoes and bookcase doors is to hold together the many pieces of furniture of varying styles and provenances. These include antiques from Queen Anne to Empire with recent Modern, including a Steinway grand piano and an eight-foot upholstered sofa of unique design, plus a smaller pillow-back type. There are fireplaces at both ends of the room, each attracting to itself a comfortable conversational grouping in which one of the sofas plays a major part.

Although most of one wall is occupied by bookcases with cupboards below, there are no less than eight doorways—three of them round-topped arches, two with triangular pediments, two flat-topped interior doors, and one wide, paneled exterior door, leading to a terrace, which has full-width strap hinges and a four-paned rectangular fanlight. The two windows are framed in straight moldings, with a plain cornice below that of the ceiling, and each archway (and therefore the fireplace) flanked by fluted pilasters, while the mantel has an eared surround over its dark-marble facing and a heavy, carved molding like those over the main-floor front windows as mantelshelf.

The most important piece here is a Queen Anne double, bonnet-topped bookcase desk in walnut dating from the seventeentwenties. There is also an early Sheraton library table in maple; a Chippendale wing chair in the same material as the curtains; and some Chippendale and Sheraton side chairs, sufficient to cater to the most eclectic taste. Finally, the stained-oak floor is partially covered by a large Oriental rug, augmented by a Baluchi runner.

It would, however, be absurd to take leave of so charming a residence without a glance at the halls, upper and lower, which emphasize the factor of hospitality for which these early New England houses were noted. For one thing, the entrance hall is far more spacious than the apparent size of the pavilion would suggest. It is also blessed with a number of elaborate architectural features, such as the eared side doorways to the front rooms; the heavy modillioned cornice common to the better rooms; fluted pilasters and a raised-panel dado that also winds up the staircase; the wide stair whose step ends are ornamented with applied jigsaw scrolls in stained wood over the white paint; and a hexagonal brass lantern to illuminate it all.

On the second floor one immediately comes face-to-face with the lovely Palladian window, looking very much as it does from the exterior. Here the wide landing with its two eared door-frames and paneled dado has a floor area more than sufficient to accommodate a pair of bookcases and a large library table with its bouillotte lamp, consorting with a tall-back carved armchair and an upholstered one, all on a rich, colorful geometric rug that completes an enviable retreat for reading or correspondence or a chat with a friend while looking out over the quiet, treelined street that this house has graced for almost 200 years.

The huge living room is a fine example of eclectic furnishing.

Savannah, Georgia

THE DAVENPORT HOUSE

A Modified Post-Revolutionary Georgian

Davenport House, tall and dignified, with its flying exterior stairways.

Isaiah Davenport was one of the few people of the Revolutionary period who had the knowledge and capacity for designing and building his own house. Furthermore, he was independent enough to select the style that would best serve his needs regardless of current fashion, while modifying each room to suit its particular purpose and endowing it with individuality, all within the canons of good taste. All of this he was able to do, at the same time achieving a maximum economy of space.

Davenport, born in Rhode Island in 1784, was of English descent, being related to the founders of the Davenport Pottery at Longport, Staffordshire, with whom he apparently maintained contact throughout his life. After serving his apprenticeship in building-design construction at New Bedford, Massachusetts, he moved to Savannah at the turn of the century and quickly made a name for himself in that field. In 1809 he married and three years later bought land on which to erect a town house in keeping with the needs of his growing family and also suited to his prosperity and standing in the community. However, another eight years were to elapse before slack times, resulting from a national recession, allowed him time to proceed with the building.

His initial choice of an Early Georgian style and the changes effected as building proceeded are an interesting indication of the manner in which he selected the best features from a variety of current fashions and possible future developments, borrowing alike from Adam and the later Greek Revival vocabulary.

In any event, this sturdy red-brick house has all the presence and accommodations of a larger mansion, successfully combining a basically Georgian structure with certain neoclassic details of the Federal period. Its rectangular windows, for example, have lintels and sills of sandstone—with the exception of the exposed-basement windows, which have flat brick arches—while the main entrance door, which incorporates leaded sidelights, is topped by an elliptical fanlight that spans all three and is, in turn, crowned by a sandstone arch, a distinctly Federal feature.

Apart from these window and door details, the exterior has few decorative features, relying for its air of importance on its proportions and contrasting textures, with some help from the green-painted blinds that split the façade into a group of horizontal bands of alternating brick and painted wood that emphasize its width.

The handsome appearance of the principal façade is due in no small measure to the twin flights of steps that curve around the ends of the stoop, the iron handrails being joined at the top by a decorative black iron grille. Underneath this stoop, the front wall is pierced by a heavily barred, round-topped opening that serves to ventilate the basement, while quadrant-shaped spaces at the bottoms of the steps are planted with thriving holly bushes that add an air of distinction to the whole unit.

Visible above the mansion's high façade is the terneplate roof whose twin end chimneys are tied together by gable parapets in typical Georgian style, while three dormers with round-topped sash windows are spaced along the eaves to complete the period picture. At the rear of the mansion the entrance doorway, leading to the through hall, is covered by an attached wooden porch at garden level; otherwise the façade duplicates that of the front.

The main floor accommodates four rooms, no two of which are the same size, the original dining room, at the rear, in particular being sharply curtailed by the necessary widening of the rear hall to accommodate a U-shaped staircase that winds up to the attic, leaving a wide, open well in the center. Although the front hall is comparatively narrow, it still provides ample room for the necessary furniture, such as a John Bell tall clock, dated 1800, and side tables, one of which is an English serving table, the other an octagonal, folding English card table of which there is a near duplicate in the stair hall.

In the front hall a feeling of space is contributed by the two recessed doorways opening off it, of which that to the drawing room is double. Here the high ceiling is bordered with an effectively simple flat molding running down into the cornice, and from its center hangs a delightfully plain four-sided brass lantern having a crown-shaped crest.

At the junction of the two halls where the front hall ceiling ends are twin Ionic columns, painted white and crowned by wide sections of a simplified frieze and cornice supporting an elliptical arch—a favorite device in many an Adam and Regency interior up into the 1860's—forming a graceful introduction to the rear hall, whose end wall is curved to match the sweep of the winding stair. This hall, running alongside the enclosed stairway, under which steps lead to the basement, actually is but little narrower than the entrance hall and not only provides space for a side table and chair and a marble pedestal type of plant stand, but also allows the front

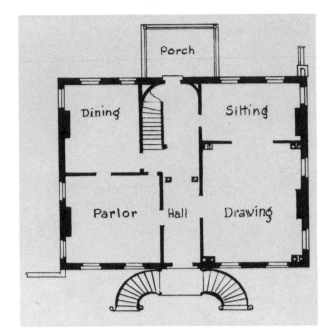

The main-floor plan features two notable flights of stairs.

Front and rear halls are joined by a pillared archway.

This lovely staircase is flooded with light by a round-topped window.

Over the Hepplewhite sideboard hangs a bullet-punctured portrait.

and rear antique rugs to be neatly aligned, breaking the sweep of the mahogany-stained pine floorboards so that the perspective is agreeably shortened. The curve in the stair also reveals part of the high rear window lighting the upper section of the stairway, adding interest to the whole interior. This round-topped window, incidentally, reaches up to the underside of the flying staircase almost at second-floor level, at which point, it is interesting to observe, the molded plaster cornice stops short on either side of the keystone of the window arch—a logical but most unusual treatment which underscores the originality of the plasterer, a certain George Smith who scratched his name in the plaster of the drawing room arch in September, 1822.

This drawing room archway itself constitutes another remarkable example of the interior architecture of the house. Under ordinary circumstances, this room would be pleasant enough with its double hall doorway, a wide, flat archway into the next room, and four large windows each enframed in a deeply molded architrave with carved corner blocks. The front pair of windows is separated by a round-topped, gold-framed pier glass; the others flank an Italian white-marble mantelpiece whose frieze is adorned with carved panels and paterae and supported by quadrangular terms with draped feminine heads surmounted by Ionic-style scrolls. To these are added a lovely French ormolu-and-crystal, eighteen-taper chandelier, surrounded by walls and woodwork of a soft gray-green, keyed to the blue and gold of an antique Indian rug from Agra and the ivory tones of damask draperies echoing those of the mantel. As if that were not enough, an unusually elaborate plaster cornice sweeps right around the room.

Such a beautifully decorated interior was apparently not sufficient for Davenport's creative urge. He added a somewhat grandiose note by installing in each corner of the room a pair of modified Ionic columns, one in the round, the other attached to the side wall, mounted on a common plinth and capped by a plain frieze and modillioned cornice as imposts for an elliptical arch extending across each end wall, a paneled block at its center, like a memorial tablet, running up into the projecting cornice.

The furniture here, though simple in design, is worthy of its setting. Included are shield-back Hepplewhite mahogany chairs and a twin-back settee in a pale-blue French silk of a coin-dot pattern; Chippendale straight-legged side chairs in antique-gold damask; and a pair of Irish Chippendale chairs with beautifully carved backs, four cabriole legs, and brass-nailed black-leather seats. There is also an eighteenth-century, Queen Anne, tilt-top tripod table on which is set out the Davenport coral-and-gold china tea set made especially as a wedding present for Isaiah and his bride.

By the fireplace is an English mahogany pole screen with hand-painted, twin parchment shields that tilt to form a candlestand, and over the mantel hangs an eighteenth-century portrait of the Reverend James Willson of Bohemia Manor, Maryland. In the corner of the room by a window is an American rope-leg side table, with a portrait of James Maury, first American consul to England, on the wall above it.

Another interior, far less elaborate but still redolent of the period, is that of the dining room which, though small, is beautifully furnished with Sheraton and Hepplewhite pieces. It also displays more Davenport china in coral and gold, together with some of the family's original silverware made by Savannah's silversmith, William Penfield, in 1827, all within a setting of warm pinkish-beige walls with white trim, maroon damask draperies, and a rich maroon Bokhara rug with a design in ivory and blue. The floor itself is of a dark pine which has a certain affinity for the mahogany pieces, while the chair seats, in gold damask, supply pleasing accents.

The white-painted wooden mantelpiece, not original to this room, is enlivened with a shelf display of Davenport china in the colorful Imari pattern and a portrait of General Charles Williams, *circa* 1850. Against the adjacent wall is a Hepplewhite sideboard made in Baltimore in 1790, now graced by a pair of old Sheffield candelabra of an exceptionally delicate design. The painting over this beautifully banded piece is one of an unidentified gentleman (after Sir Gilbert Stuart), a temporary substitute for the portrait of General Williams' wife, damaged by Yankee bullets when Sherman's men pillaged

One of the drawing room's Irish Chippendale side chairs.

The dining room features a 1790 Hepplewhite sideboard and Sheraton table.

A master bedroom feature is a rare folding fireside desk.

The southeast corner of the drawing room.

the Williams' plantation house near Gray, Georgia, during the Civil War and removed for repair.

In this room the small dining table, set for use, is the center section of a three-part Sheraton extensible unit. This occupies the middle of the floor beneath a crystal chandelier similar to that in the drawing room, but minus the ceiling rosette which was past repair when the equally intricate cornice was restored. The rounded end sections of the table are used as separate side tables, displaying china in the window recesses alongside the fireplace.

In the sleeping quarters, the architectural details are of comparatively simple design, as a brief examination of the master bedroom discloses. Here the cornice molding is strictly geometrical though far from simple; the door trim is almost plain, but the fireplace mantel is more elaborate and echoes the drawing room woodwork with reeded uprights framed by three-quarter rounds and a frieze panel with a recessed, beaded oval in its center.

This room has the same stained pine floor as the other interiors, a dark tone that makes an excellent background for the antique Persian rugs in ivory, crimson, and blue. In this quiet setting the window draperies and the four-poster-bed hangings in a wine-red *toile de Jouy* look their gay best; their only competition is a Chippendale wing chair in red and gold silk damask. The bed itself is an exceptionally fine piece made in Charleston, *circa* 1790, with rice carvings on the vase-based, fluted mahogany posts.

By the fireplace there is a folding fireside-screen desk of mahogany from England, dated around 1800, and alongside the wing chair a kettle stand serves to hold a candlestick, while a round-topped tripod table at the bedside bears a *tôle* candle-stand with an adjustable shade. The rest of the candle holders, including a pair on the mantel, are of bell brass in keeping with the singularly dainty fireplace accessories, the brass-rimmed, iron-mesh fender, and the gilt frame of the overmantel portrait of a young English girl of the early eighteenth century.

Isaiah Davenport died in 1827, only 5 years after he had taken up residence in the house, and his wife and family moved away in 1840. In 1955 the dilapidated mansion was rescued from total destruction, restored and refurnished by the Historic Savannah Foundation, organized for the purpose. Fortunately the original inventory of the Davenport possessions was discovered, and this was used as a guide in acquiring similar furnishings either by gift or loan, so that today the mansion appears much as it must have when occupied by its builder and his family 150 years ago.

OPPOSITE: This sitting room features an early map of the Savannah waterfront over a typical "broomstick" mantel.

Nashville, Tennessee

BRAEBURN

A Twentieth-Century Educator's Dream House

The row of giant Corinthian columns forming Braeburn's front portico evidences the classical taste of the two spinsters—Miss Ida Hood and Miss Susan Heron—who designed it as their retirement home in relinquishing possession of Belmont College for young women, as they did in 1912.

Its huge drawing room, 46 feet long, likewise bears witness to their determination not to lose all contact with their past, but to surround themselves with throngs of young people, as they had for so long, while a small French parlor with cupids disporting themselves on the ceiling betrayed their interest in mythology, romance, and the art of the Italian *Risorgimento*.

Among the more practical aspects of the interior design was the specifying of identical bedroom suites for the two of them so that neither would ever have cause for dissatisfaction or envy. Meanwhile the attic space was to be reserved for a ballroom to be completed sometime in the indefinite future, while the basement was to be devoted to the mundane domestic activities and the servants' quarters.

This, then, was the dream house that was to materialize in 1914, thanks to a competent builder (no architect being involved) who was given a free hand in interpreting their desires while they scoured Europe for treasures with which to furnish it. The only major setback to the building of the house was due to the unexpected arrival of a semicircular staircase from France and the structural alterations that had to be made to accommodate it. In contrast, the only deviations from specifications made by the builder were minor. The owners had asked for a Greek temple-style house of white-painted brick, and what they got was a peak-roofed house of a warm-gray limestone—of local origin—fronted by a classic portico whose flat roof was tucked under the eaves and adorned with a decorative balustrade that concealed three pedimented dormers but not the gable overhangs or the tall end chimneys.

Set on a foundation six steps up from the laterally sloping front lawn, the mansion was both stately and impressive with

Braeburn with its classical Corinthian portico.

OPPOSITE: In the dining room a hand-blocked Zuber scenic paper competes for attention with the exquisite Baccarat chandelier.

Floor plan of Braeburn today.

The surprise semicircular staircase from France.

its classic portico, giant end pilasters, and flanking square-headed sash and ground-floor French windows, all grouped around a handsome pillared doorway with sidelights and transoms, and a wooden balcony, paneled and bracketed in white-painted iron filigree, accessible from the second-floor hall. Though the entablature above the columns is quite plain, the balustrade is divided into openwork geometrical panels, exquisite in design and proportion, which adds both character and beauty to the main façade.

The original owners, being of Scottish descent, gave the house the nostalgic name »Braeburn.« Set within ten acres of tree-dotted lawn, once a deer park, the structure is quite simple in plan, though impressive in its interior dimensions, featuring a front-to-rear hall 48 feet long and 12 feet wide, off which opened the French parlor to one side and the 46-foot-long drawing room to the other. A side hall extended to the dining room, and a semicircular recess accommodated the famous stair, the whole terminating in a wide doorway to the breakfast room.

In 1940 the house was acquired by Caleb Haun, whose family was to occupy it for the next twenty-four years. During that time a glass-walled sitting room was created by extending an existing palm room or solarium, minus the roof balustrade. Behind this was laid a patio paved in speckled concrete tiles, accessible from the sitting room, the drawing room, and the breakfast room. Meanwhile the little French parlor had had its gaudy ceiling painted white, and its walls encased in butternut paneling as a more fitting decor for its new role as the library.

In 1964 the mansion was acquired by Vanderbilt University

Braeburn

At the rear is this tile-paved patio.

The glassed-in Green Room adds a modern touch.

Antiques enliven even the
breakfast room.

as residence for its chancellor, Alexander Heard, and his
family whose furnishings, including many heirlooms going
back a century or more, now augment certain basic pieces
already installed, including a delicately beautiful Louis XV
hall lantern in gilt bronze, now suspended at main-floor ceiling
level in the stairwell.

Actually this great hall constitutes an attractive room in
itself, with the extra advantage of providing interesting vistas
through the wide doorways to the drawing room whose double
sliding doors are ordinarily open, and that developing as one
approaches the semicircular sweep of the widened hall, where
the stair curves downward to end in a cluster of white-painted
balusters on the round-nosed curtail step. Equally impressive
are the architectural features in the form of paneled pilasters,
heavy, molded ceiling beams, and boxed cornices, each of the
deep, cofferlike recesses illuminated by a crystal-draped
ceiling light at its center.

All of the paint work here is off-white, the walls being covered
with a cream-colored paper simulating a damask pattern, and
the fine oak floor almost hidden by a wide, modern piled
runner in dark green. All of these combine to emphasize the

This central hall is 12 feet wide.

A side hall leads to guest quarters.

A feature of the main hall is this antique breakfront.

The hall's American Empire side table and Chippendale mirror.

A Régence-Louis XV gilded
canapé and *tables en chiffonnières*
also grace the hall.

The library features the curio
cabinet.

attractive features of the furniture of mixed periods and varying styles. Just how well these unalike pieces can be combined is illustrated by the groupings spaced along the walls, separated by the pilasters, doorways, and passages. These groups actually invest the hall with an atmosphere of elegance, regardless of the angle from which it is observed. Close by the front door, for instance, a brass-footed American Empire side table is set beneath a carved and gilded Rococo mirror, a fine Chippendale design *circa* 1760, combining leaves and C-scrolls with floral drops, the two pieces tied together visually by a conical floral display flanked by a pair of brass candlesticks.

Across the hall from this is another group composed of a Régence/Louis XV transitional gilded canapé in pale-green damask, on either side of which is a small marquetry *table à ouvrage* with a marble top and bronze gallery, one of them serving as a lampstand. Centered above these is an heirloom painting of a Danzig guildhall interior, artist unknown, while in front of the canapé stands a wrought-iron coffee table with a top of rose-tinted marble.

Between this group and the second doorway to the drawing room is an 8-foot-long, nineteenth-century English breakfront bookcase of mahogany with ebony trim, displaying blue Canton china used by four generations of the Heard family. This happens to face the lovely semicircular staircase nestling in its rounded recess with only the curved bottom step projecting. Almost directly under the hanging lantern is a great oval pedestal table of walnut—a modern but exquisite reproduction of an earlier piece made in Nashville. Almost always this tabletop is occupied by an enormous bouquet of sea oats and eucalyptus in a footed brass bowl, sharing the space with an oak case of English dueling pistols once the property of Mr. Heard's great-grandfather, Thomas Butler King. Just beyond this point the visitor comes face-to-face with the glass-paned doors opening into the breakfast room.

In addition to its unusual size, the drawing room can boast of an extraordinarily compatible selection of furniture pieces, old and new, set against a background of faintly green walls and ceiling and a floor wholly covered with a dark-green broadloom carpet. This color combination brings out the variegated hues and graining of the furniture woods, from the dark walnut of a pair of beautifully carved Italian armchairs in gold damask to the parcel-gilded Adam pieces. Whether by accident or design, each furniture item in this room not only is lovely in itself but also helps to emphasize the fine qualities of the rest. The Adam chairs, for example, with their semicircular caned backs incorporating a medallion with gold cupids on a green ground contrast beautifully with the small and delicate pair of ormolu-mounted and inlaid chiffoniers as well as the gilded Louis XV banquette in apricot velvet.

The white damask sofa is, of course, modern, as is the long coffee table with its white-marble top, its pierced and carved apron inset with gold and accented with carved fleurs-de-lis,

A corner of the library.

yct both tied in so perfectly with the twin Louis XV-style *fauteuils* in their green and white-striped damask. Squares and curves also are acceptably combined in the two tables flanking the sofa—one a Hepplewhite console with reeded legs, somewhat faded with age; the other a Victorian drop-leaf table with carved underpinnings, both in nicely grained mahogany.

Drawn up to the hearth is a pair of Louis XVI *bergères* in cinnamon velvet, with another of the exposed-frame type covered in green damask, placed in the nearby bookshelf corner. The rather plain mantelpiece is of green-painted wood with a facing of white marble, a perfect setting for the garniture of two red-and-white lidded Sèvres vases touched with gold, which were bought in France in 1808. In between them hangs a gold-framed portrait of Thomas Butler King.

Other important pieces in this room include a carved fruitwood cabinet of Italian design with a marble top and a pair of painted Adam chairs with oval caned backs and cane seats. Completing this group is a French ormolu shelf clock, with a ruby and gold Venetian-glass vase on either side and a Piranesi print on the wall behind it.

Even more formal than the drawing room is the dining room, particularly on special occasions. At such times the display of English silverware and Wedgwood china in a setting that includes off-white and gray Zuber wallpaper, pale-gray walls, and darker dado contrasted with draperies in gold-fringed, off-white antique satin and a brilliant Oriental rug in red, blue, and beige is little short of overwhelming.

The airy guest room is also furnished with antiques.

An interesting drawing room grouping.

The trim, including the twin-leaf folding doors, is painted white, and from the pale-gray ceiling hangs a tremendous Baccarat chandelier with branches of gray glass, its forest of sparkling white crystals reflected in the gleaming mahogany top of a three-part banquet table which sports carved and rope-turned legs. Close by is an inlaid Hepplewhite-style tambour sideboard laden with a sterling tea set from England and twin candelabra whose tapers, like those of the dining table, are a pale yellow, called autumn gold, which matches the seat fabric of the chairs ranged around it. Auxiliary lighting is supplied by delicately crafted twin-candle wall sconces of a grape-leaf design in gilt bronze from Italy.

Across the hall from the front end of the drawing room is the library which, in spite of its comparatively small size, houses many treasures. It is unusually cozy with its butternut paneling and fireplace surround, its draperies of saffron silk, the whole enlivened by a gaily patterned Sarouk rug and the Louis XV side chair, Boulard style, in assorted tones of brown, plus a wing chair whose figured linen upholstery contributes masses of apricot, brown, and beige.

Flanking the fireplace are two remarkable three-legged chairs, intended for Spanish shepherds, which add an exotic touch, counterbalanced somewhat by shelves of books in plain and fancy bindings and many a paper jacket. Other contrasts are supplied by a cream-colored sofa accompanied by a mahogany butler's tray serving as coffee table and by a reproduction of an old walnut curio cabinet full of bibelots, accompanied by a Seignouret type of mahogany chair, in Empire style, reputedly shaped in one piece from a single mahogany log by Mr. Heard's father in 1940. Altogether, these are indeed surroundings well calculated to induce a contemplative mood.

OPPOSITE: This end of the eclectic drawing room features Adam, Italian, and Louis XV chairs, and inlaid French commodes.

THE WILLIAM GIBBES HOUSE

A Revolutionary Georgian Classic

One of the country's finest Georgian town houses, and certainly among the most beautiful insofar as classical detail is concerned, is the William Gibbes House in Charleston, South Carolina. It is also one of the most individualistic. Now the property of Mr. and Mrs. Ashby Farrow, it was built by Gibbes in 1772, though certain important changes were made in 1800, thus straddling the Revolutionary period during which it became subject to Adam and Federal influences as well as the original Palladian, all of which have left their mark.

Although its exterior walls are of white-painted, beaded clapboard of black cypress instead of stone or brick, it has all the character of its basic Georgian style, together with a certain intimate charm with which, theoretically, the more formal and enduring materials could not have invested it. It also owes something to its imposing height, thanks to the raised basement which is almost wholly aboveground, a characteristic of many mansions in that area, and of particular value in this instance where the house actually faced upon the water and was therefore subject to possible flooding. In those days the Ashley River was no farther away than the width of the road, and since Mr. Gibbes was a ship owner, he apparently was influenced in his choice of a site by the fact that this would permit him to erect his own wharf opposite the house, as he later did.

In any event, the high basement was built of random stone, stuccoed over and rusticated with V-joints so that the cement would look like ashlar. This treatment was extended to the front wall of a basement jog which supports the short terrace fronting the main doorway of the house. In the center of this projecting wall, at street level, is the arch-topped service entrance which is flanked by sidelights imitating, on a smaller scale, those of the mansion's main entrance. At each end of this basement-level structure, iron-railed marble steps lead up to the terrace. At their center the terrace rails incorporate both a decorative ironwork panel and a pair of tall, wrought-iron lamp standards with glass globes.

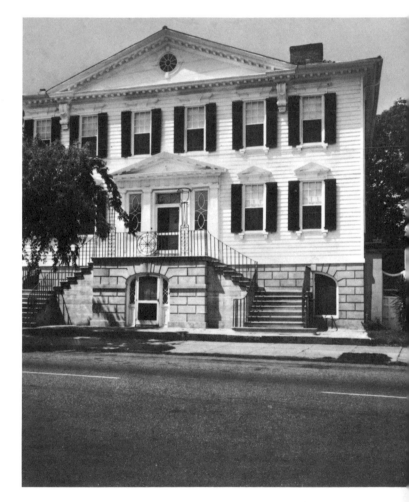

The house Gibbes built in 1772.

OPPOSITE: The ballroom is pure Adam with its coved ceiling and mantel, setting of a huge Louis XVI vitrine.

45

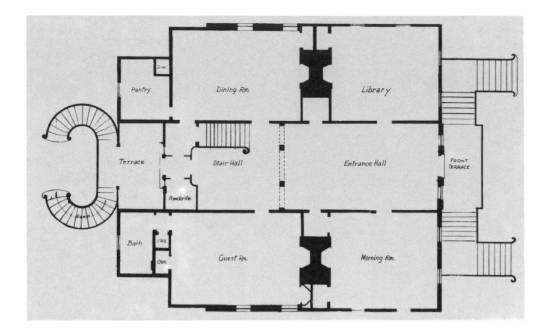

Floor plan of the house today.

The rear of the house from the garden.

This handsome central grouping of steps, service entrance, and platform is a fitting complement to the structure above it which has a number of splendid features not altogether Georgian. For one thing, the main entrance unit expands the importance of the comparatively narrow doorway by flanking it with two pairs of pilasters, and placing between each pair a broad, decorative window admitting light into the wide entrance hall.

Spanning four of the six top-floor windows is a massive roof pediment, much like a cross gable, supported at each end by elaborately carved consoles, its whole length punctuated by tongue-shaped modillions which are carried right around the roof cornice. In the center of the pediment's tympanum is a simple ten-spoke wheel window serving to light the attic. This almost duplicates the pattern of the circular panel in the iron railing below. A further elegant touch is provided by the small *cyma recta* window brackets, the light and airy main-floor window pediments and their architraves with projecting centers, as well as the stepped cornices of the upper ones—small details that endow the whole front with its special character, adding liveliness to what could very well have been formal and stodgy.

Pretty much the same thing happens at the rear of the building where, in the 1920's, a small room, one story high above the basement, was added to each side, forming a recessed porch between them. From this porch, which is open to the sky, a graceful twin flight of brick steps descends in sweeping curves like the letter *C*, to the garden level. Not to be overlooked, however, is the beautifully detailed Palladian window at second-floor level, which, while adding grace to the whole structure, also lights, as well as adorns, the interior stairway and upper landing.

The entrance hall of the Gibbes House in no way belies the promise of the exterior, being itself replete with architectural features that serve a purpose besides being highly ornamental. One of these is the formal archway composed of a transverse ceiling beam supported by two marbleized cypress columns and their accompanying pilasters which divide it, visually, into a reception area—or foyer—and a rear stair hall. Actually, these two areas have little in common since the rear section is comparatively informal, housing the main staircase whose two flights are connected by an iron-railed platform across the rear wall. This short landing permits a view through a rear window looking out onto the tall treetops that shade the garden. Below it is the paneled rear door leading from the stair hall to the elevated porch.

In both the front and the stair halls, the high, white-painted dado is interrupted by exquisitely designed doors and doorways with heavily molded trim, the frames crossetted and surmounted by overdoors with paterae and dentiled cornices topped by a deep cavetto molding—all later Adam touches.

Here every white-painted surface—doors, dadoes, pilasters—stands out strongly against the dramatic, scenic, colorful wallpaper with its classical ruins, peacocks, and tall trees against a background of forests, temples, monuments, shaggy palms, and craggy hillsides—a reproduction of the nineteenth-century Zuber "El Dorado" pattern that converts the upper walls to windows on a riotous, exotic landscape. Here, too, the old pine floor sets off fine Oriental rugs in both halls, the width of the twin marbleized columns.

It is here that one gets a first glimpse of the furnishings. There are some exquisite antiques, some not so rare, all selected in their placing for utility rather than looks or degree of formality, the hallmark of a lived-in home.

In the front hall, therefore, a nicely carved Empire sofa offers comfortable seating opposite a six-legged mahogany side table in the Adam style, adorned with a pair of bronze candelabra with a bronze animal group on a marble base between them. Farther along, a nice little Pembroke table with cross stretchers faces, across the floor, a half-round Sheraton card table, while near the doors is a pair of inlaid consoles in light mahogany and twin five-banister shield-back Hepplewhite chairs upholstered in Burgundy mohair. In the stair hall there is also a nice Coromandel screen and an eighteenth-century tall case clock. The total effect, illuminated by porcelain-based lamps and twin crystal-draped lanterns, is one of easy charm.

The wide staircase has a walnut handrail with decorated wrought-iron balusters, terminating at the bottom in a sweeping curve whose center displays that quaint conceit, an ivory "peace button." The stringers, risers, and paneled partition enclosing the stair are painted white in happy contrast with the walnut treads.

One of the most interesting of the interiors is a large upstairs chamber which functions both as a music room and a ballroom. Basically, it is a beautifully designed apartment, tranquil in its pastel tints, owing nothing to violent contrast or strong colors.

Scenic wallpaper makes the entrance hall seem even larger.

The upper hall illuminated and adorned by a Palladian window.

47

One corner of the Gibbes library.

Chinese details of the library's
fireplace end.

Its high, coved ceiling betrays the Adam influence in its plaster rosettes and the fan tracery in the corners, which are heightened by touches of gold, and is climaxed by the sparkling luminosity of an ethereal chandelier with slender, curving gold branches and a myriad of finely cut crystals.

The room's excellent proportions are emphasized to some degree by the painted paneling of the dado and the upper walls, which are tinted a warm gray in contrast with the white ceiling and the delicate beige of a room-sized Aubusson rug, which is bordered in pink and patterned in tan, blue, and gold. Heavier gold accents are introduced by the damask draperies, and milder ones by the polished brass of the fireplace accessories, a French Empire card table with its eagle pedestal and dolphin feet (reputedly once the property of Napoleon Bonaparte), the gold-leaf frames on the Chinese Chippendale mirrors, and the family portraits, one by Thomas Sully (1783–1872), the other by Rembrandt Peale (1778–1860).

The Adam-style fireplace mantel, colored to match the walls, is itself a work of art with its finely carved frieze and pilasters plus a narrow facing of pink marble. On the mantel is a pair of quite unusual miniature paintings on ivory by Sully. At this end of the room are the musical instruments—an elaborately decorated Venetian harpsichord with painted panels (said to have been played by the immortal Ignacy Jan Paderewski) and a French harp made by Sébastien Érard (1752–1831), together with the tripod stool and tall, spidery music stand, completing a picture of dignified elegance as suited to 1772 as to tomorrow.

Even the larger furniture contributes to the delicate atmosphere, most of the pieces being of the Louis XV and XVI styles,

48

the earlier cane-back chairs painted a pale blue-gray and gold, with blue-gray cushions, while the later ones preserve their natural applewood finish and are upholstered in a pale rose and blue brocade to match the original fabrics still in use on the Louis XVI sofa. The largest and most impressive piece of furniture in this room is a Louis XVI breakfront bookcase, once painted and now bleached. This serves not only to display fine bindings but also antique porcelains, glassware, and fans. Standing next to the fireplace, it occupies most of one wall and adds largely to the dignified character of the whole interior.

A somewhat smaller room, but one with its full quota of charm, is the library, whose original Georgian details were changed to create a Chinese setting by a former owner (Mr. Farrow's grandmother), Mrs. Washington G. Roebling, who collected Oriental artifacts. Though the room is on the main floor, the white ceiling is coved; from it is suspended a four-light crystal chandelier that once adorned another Charleston house. A deeply coved and reeded cornice tops the paneled walls, all except the fireplace wall having a carved chair rail and dado. It is the fireplace wall, however, that contributes most to the feeling of elegance of this interior with its beautifully carved overdoors and mantel, all three of which carry the same acanthus motif.

The overmantel panel is occupied by a large framed portrait of the Chinese warlord Kwang Gnu, with triple crystal wall sconces on either side of it, while the mantelshelf is garnished with gaily colored Chinese funerary figures dating from the Tang dynasty. The rest of the walls are covered with English wallpaper panels of Chinese design having a floral pattern in

A peep into the morning room.

A notable Queen Anne period writing desk with vitrine top at the opposite end of the morning room.

The Bishop's Room.

A stunning bookcase in
the ballroom.

shades of pink and tan on a pale-green ground. All the wood-work below these is painted a pale green to match the fireplace wall, and the draperies are of an iridescent pink and green taffeta with gold and green tassels, most of these tints being picked up by the solidly patterned Oriental rug.

Among the furnishings are three upholstered pieces in gray-green damask and a pair of Chippendale chairs with reeded stiles covered in red velvet, which obviously supplies an agreeable accent. There is also an especially fine shield-back desk chair of Charleston provenance in gold-striped silk damask, with beaded legs, arm grips in an inlaid wheel pattern, and an oval plaque inlaid with flowers over the carved feathers in the back. But, best of all, is the Chippendale mahogany secretary-bookcase with blind fretwork around the frieze.

A really good example of fine furnishings for everyday living is provided by the Gibbes House morning room, an informal interior designed for relaxation and for casual family use without the slightest concession to quality in the room's appointments. Here again the paneled walls are painted a gray-green, together with the dado which extends into the recessed window bottoms and abuts against the fireplace mantel.

This white ceiling, though flat, has its Waterford-crystal chandelier (also from an older Charleston house), and the floor its Oriental rug, with pink and green draperies exactly as in the Chinese Room. The fireplace, however, differs considerably, being of white marble with carved and reeded pilasters and pink marble insets on either side of the carved central panel. In the carved and crossetted overmantel panel is a large, early nineteenth-century, bull's-eye convex mirror with its ball ornaments, horns of plenty, and flower basket as crest.

In one corner of the room is a lovely Charleston-made cupboard with a broken pediment and urn finial, its vitrine doors having octagonal muntins, behind which are displayed historic porcelains and Oriental Export chinaware. Of almost equal interest are the eighteenth-century Queen Anne lowboys flanking the fireplace, while against an adjacent wall is a small but exquisite walnut desk of the Queen Anne period with a bonneted vitrine full of Meissen figurines and similar pieces.

In the opposite corner stands an eighteenth-century fall-front desk in inlaid mahogany, its Empire saber-legged chair looking far less out of place with its Sheraton and Hepplewhite opposite numbers than might be expected. The colors, too, are interesting, pairing a sofa in a pattern of black and white birds on a pale-pink ground with chairs in beige satin damask and multicolored petit point, details that emphasize the eclectic nature of the furnishings—common to the rest of the rooms—which characterize the Gibbes House.

Front end of the upper hall
looking into the music room.

New Orleans, Louisiana

THE FRANK G. STRACHAN HOUSE

A Southern Classical Masterpiece

A lovely mansion in delightful surroundings perfectly describes the Frank G. Strachan house (pronounced *Strawn*), which combines good architecture in the neoclassical style of the mid-nineteenth century with equally fine landscaping that is constantly being improved upon so that it still remains, after nearly 120 years, one of the showplaces of the New Orleans Garden District.

The house was built by Jacob U. Payne, a nearby plantation owner who bought the land in 1849 and immediately began work on the mansion which, it is assumed, he also designed, though doubtless borrowing many a suggestion from Minard Lafever's *Modern Builders Guide*. Remaining the property of the builder's descendants for the next two generations, the mansion was sold in 1935 to Mr. and Mrs. William Bradish Forsyth, who later willed it to their daughter, Mrs. Frank G. Strachan, into whose possession it came in 1940.

From the front, the white-painted, stuccoed brick residence is wholly classical in appearance, with its Corinthian-columned balcony supported by the main portico of Ionic columns, all of which, incidentally, have cast-iron capitals marked "N.Y. 1848." Above these, a deep entablature with a low parapet hides the ridge roof behind it—a roof which terminates in parapeted gables that tie each pair of end chimneys together well above the ridge, a practice dating to the early 1700's that adds much to the appearance of massive solidity. At the rear the two galleries and entablature are repeated, though the classic columns are replaced by heavy, square posts, and the upper balcony has a cast-iron railing of a somewhat different pattern from those at the front.

Another decorative feature—a Strachan addition—is the two-story, cast-iron balcony structure against the south gable wall, enshrouding two pairs of windows. Standing on its own flagstone terrace, this is flanked by Confederate-jasmine, vine-covered iron trellises, while above its roof are the two third-floor gable windows. Together with its box-bordered garden

The classical façade and curbside monument to Jefferson Davis, who died here.

The Frank G. Strachan House

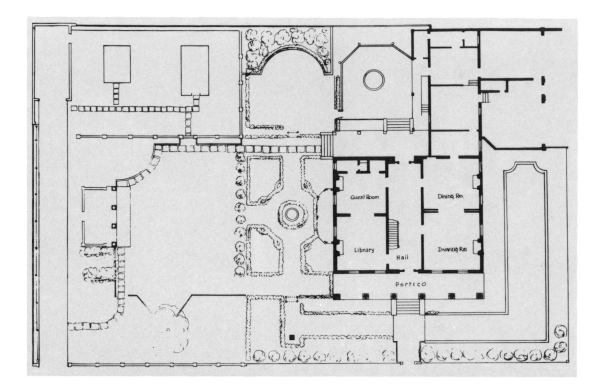

Plan of house and gardens.

A garden view emphasizes the lovely Southern ironwork.

plot, this makes a delightful ensemble overlooking the formal garden with its parterres, arches, vistas, and embowered statuary—all in formal Italian style, as planned by Umberto Innocenti of New York—terminated to the south by tall cherry laurel bushes, and redolent of night-blooming jasmine, crepe myrtle, and oleanders. Under those tall trees nestles an open-fronted classical pavilion, with slender Ionic columns and an urn-bedecked parapet, all a gleaming white—and far more recent than it looks.

From the roadway, the main façade with its columns, iron tracery and tall sash windows, the massively classical eared surround of its paneled door and narrow sidelights, all spanned by a six-step-high porch, and flanked by standards of ivy in green-painted tubs, is a joy to behold though half-concealed by the heavy gate pillars and tall iron-railed hedge of podacarpus. Even its sidewalk grass plot set with flagstone squares forms a delightful proscenium that foretells of equally inviting enchantments within.

Inside, the entrance hall is rich in an unpretentious manner—the walls of Wedgwood blue, the trim white; the woodwork carrying out the classic theme with high, eared, and angular overdoors, monumentally beautiful instead of merely pretty; the doors of mahogany, and the dark floors of heart pine; the long, straight stair climbing through the high ceiling above a molded frieze adorned with a line of paterae, a detail repeated throughout the principal rooms. This simple theme is picked up by a huge ceiling rose from which hangs a sparkling brass lantern to light up the front hall and stairs, the latter painted white in contrast with the mahogany steps, handrail, and turned balusters; the steps and risers are covered in wide, gray carpeting with a silky sheen.

OPPOSITE TOP: A mid-nineteenth-century painting of the house in which Jefferson Davis died.
OPPOSITE BOTTOM: The drawing room features Parisian pier glass duplicating that of the dining room.

Eleven thirty four First Street New Orleans

By the front door is a Simon Willard mahogany tall clock with quarter columns, dating from the 1790's, and, facing the stair, a beautifully inlaid marquetry commode of Louis XV design. Above this is a carved and molded, gilt-framed English pier glass which is flanked by an exquisite pair of four-candle English-crystal sconces that light up the rear hall, showing off the carved oak chest beneath the stairs and the double settee in dark-blue velvet across the way. Alongside the stairway extends a long hall runner in a diamond pattern, while the rear hall is enlivened by a wider antique Sarouk rug in dark blue, red, and gold.

On entering the hall, the first door to the right opens into the drawing room, resplendent with walls of bois de rose whose frequent pilasters and high baseboard are a shade darker; the frieze and ceiling are white. The huge windows which, like the pilasters, are frieze-high, are enframed by deeply pelmeted draperies of faded silk damask with a floral pattern in deep rose and turquoise on a beige ground, over cream-colored sheer glass curtains. Against this mass of color the mantelpiece, in variegated black marble, stands out boldly, as does the cornice-high pier glass in its massive gilt frame, made in Paris for this house in 1848.

On the dark pine floor is a huge Indian rug from Agra, dotted with countless leaves and flowers in pink and brown on a beige ground, while over its center hangs an exquisite Louis XVI chandelier dated 1790, whose Baccarat crystals are mostly in the shape of teardrops, the curving arms and finial in bronze *doré*. Between two windows stands a tall, impressive Empire secretary with three urn finials, both upper and lower sections painted to represent inlay, the vitrine top displaying antique fans, Chelsea figurines, a Capo di Monte urn which belonged to Jefferson Davis, and some Venetian glass.

To the right of the fireplace is a lifelike portrait by an unknown artist of Mary Caroline Robert, direct descendant of Pierre Robert, first Huguenot minister in Charleston, South Carolina, and great-grandmother of Mr. Strachan. Under this stands an inlaid walnut cabinet decorated in bronze, with rounded, glass-paneled ends, its provenance Scotland, though, curiously enough, sometimes referred to as the India cabinet.

Other interesting pieces are a Spanish seventeenth-century gilded terra-cotta madonna wearing a silver halo on this cabinet; an early French Provincial wing chair; and a high-backed armchair, in highly contrasting upholstery materials. There is also a beautiful Louis XV serpentine, three-drawer commode in carved oak. Above this hangs a pair of colorful Aubusson tapestry panels in the slimmest of gold frames, and garnishing the top, a pair of fine eighteenth-century celadon Ming platters.

Beyond this room's sliding doors is the dining room, whose walls are covered in a white flock paper, its diaper pattern reproducing that of the silk-damask pelmet; the Scalamandre draperies which hang within the pilaster-flanked reveals are

A narrow but impressive entrance hall.

The hall from the rear.

OPPOSITE TOP: Tastefully assorted Italian and French pieces adorn the daughter's room.
OPPOSITE BOTTOM: The guest room with its muted color combinations.

55

The dining room displays a Pickersgill portrait and a tiny Corot.

oyster white. The trim also is white, the total effect being that of a room flooded with daylight. The Sarouk rug is a deep rose and dark blue, with corners of a greenish-blue tinge, which echoes the richness of the Waterford chandelier (*circa* 1780) with branches of rope-twisted glass, beneath which swags of tiny crystals suspend large ones of the fingerprint type. Over the black-marble mantel is another Parisian mirror with a carved and gilded frame, duplicating that in the drawing room.

In one corner of this room is a recessed, round-topped, illuminated china cabinet set into the wall, and in this are displayed initialed Sèvres plates made for the use of Louis-Philippe at Fontainebleau, together with rare Limoges cups and saucers. On the wall nearby is a large portrait in oils of Mrs. Strachan's great-grandfather, William Forsyth, QC (1812 – 1899), painted by William Pickersgill, RA (1782–1875), while over a serving table is a small but typical Corot landscape dated 1873, two years before his death.

In such a setting the mahogany furniture looks its handsome best. The table is an American piece with birdcage pedestals; the chairs are English Chippendale with flat crest rails, straight legs, and seats in pale-green damask. Under the portrait is an English Sheraton sideboard (*circa* 1770), with its handsome sterling-silver wine coolers, dated about 1800, a 1780 coffee urn, and a pair of silver candelabra of the same period. The silver used on the dining table is equally notable, including a soup tureen and a pair of sauceboats made in London in 1743, together with four silver candlesticks of the Late Georgian era. To one side of the pantry doorway is a large American, flat-topped secretary with shell-pattern inlays in the cupboard doors, a fitting repository for silverware and napery reserved for special occasions.

Across the front hall is the library, whose sliding doors to the guest room are permanently closed thanks to a huge, flat-topped, English breakfront secretary-bookcase set within that recess. This is a striking room from whichever angle it is viewed. Directly opposite the doorway from the hall is a black-marble fireplace which one may not immediately observe because of the impact of an even taller gilt-framed portrait of Mrs. Strachan above it, painted by Leon Kroll, which stands out against the pale-green wall. The height of this room is empha-sized by the darker green of the many flat pilasters under the white frieze and ceiling, and the tall windows draped from frieze to floor in hand-blocked Italian linen patterned in burnt orange. The floor is given weight by a large Ushak rug in gold, beige, orange, and chartreuse.

Flanking the fireplace is an English mid-eighteenth-century sofa, facing a pair of high-backed armchairs of the same prov-enance and period, all in pink silk brocade, with a black-marble-topped coffee table between them. A black-lacquered, brass-trimmed Chinese chest between the chairs serves as a smoking table. Against the opposite wall is an inlaid mahogany round-ended sofa table, over which hangs a portrait of Mr.

The garden end with its urn-bedecked classical pavilion.

Another garden vista with box hedges and embowered statue.

57

Dominating the library is a portrait of
Mrs. Strachan by Leon Kroll.

A fine Louis XV commode beneath a pair of
Aubusson tapestry panels.

Strachan by Cyprien Boulet, with a Chippendale chair to one side and an Italian painted Barochetto-style chair on the other. Interesting *objets d'art* consist of a Carl Milles head in bronze on the mantel and a modeled boy's head on a side table, a walnut square-topped tip table in the window corner, and a colorful, beaded pole screen by the fireplace, completing a picture of elegant comfort.

Beyond the breakfront-sealed sliding doors is the guest room, which is well worth examining, not only because of the richness of its decoration and furnishing, but also on account of the extraordinary contrast it reveals when compared with one of the family bedrooms upstairs. This guest room looks precisely as one would expect to find in a mansion of this quality dating from the early nineteenth century, whereas a similar interior of this same house can lend itself just as well to an entirely different atmosphere—foreign, perhaps even exotic, and in some respects quite modern, without the slightest loss of charm.

In this guest room the basic colors are *bois de rose* and beige with a scattering of blue supplied by the dark pattern of a very fine Sarouk rug and a tea set of Dr. Wall Worcester chinaware of an equally dark-blue design on white, displayed on a three-stage tripod table or dumbwaiter. The beige walls are emphasized by the white trim including the frieze and ceiling. Large areas of *bois de rose* are contributed by the draperies and bed hangings and coverlets of silk damask, plus the seats of the Hepplewhite chairs—a color that harmonizes wonderfully with the dark mahogany of the furniture. Other large neutral and color masses are supplied by a carved and gilt overmantel mirror and a black-marble fireplace whose opening is enlivened by a gold-trimmed green fan.

58

The daughter's room, on the other hand, though making use of pink draperies, window, and lamp shades against pale-green walls and rug, the latter accented with small Kashmiri rugs patterned in pink floral designs, is entirely different in character. For one thing the bedspreads are in a celadon-green taffeta; the circular table is draped in a green-velvet cover; and the furniture is of a delicately lovely Italian design, painted pale-green with gilded outlines and Rococo borders around floral bouquets, except for a Louis XV armchair in white and gold with pink flowers on a beige ground and a modern upholstered armchair in green.

Above the beds are two eighteenth-century Italian landscapes by an unknown artist, and between them a two-drawer night table. Under one window is a dressing table and stool, and flanking another are a bow-fronted secretary-bookcase and an equally lovely commode with its own oval, gold-framed looking glass with a ribbon-style crest—a finishing touch to the room that echoes the promise of the house itself: that true beauty is independent of period and therefore timeless.

The daughter's room with its delicate Italian pieces and striking eighteenth-century landscapes.

A portrait of Mr. Strachan by Cyprien Boulet occupies another wall of the library.

Rochester, Minnesota

MAYOWOOD

Home of a Dynasty of Doctors

An aerial view of Mayowood
in the fall.

The side entrance to the mansion
saves climbing many steps.

Few great houses designed by their owner and built to his specifications have been so successful in every respect as the mansion erected for Dr. Charles H. Mayo at Rochester, Minnesota, in 1910–1912. The reason behind this triumph of theory over practicality is that Mayowood, as it is called, was designed from the inside out, with little regard for either period style or current building practice.

First came the decisions as to what rooms were needed, their approximate dimensions, and the ideal relationship between them for the utmost convenience, comfort, and utility. In developing the exterior, Dr. Mayo favored no architectural type or period. The construction, too, was novel at the time, the walls, at his insistence, being formed of concrete, incorporating an air space for insulation against heat, cold, and dampness. The main walls, as a result, are eighteen inches thick, which means that all doors and windows are set in deep reveals, usually paneled, which adds to their impressiveness.

The mansion was built on a rugged Minnesota hillside and for the next fifty-five years proved to be almost ideal as a home, first for the doctor's own family, and later for that of his older son, Charles William Mayo, M.D., who before his death turned it over to the Olmsted County Historical Society for preservation as a monument to the Mayo family, founders of the world-famed Mayo Clinic and the Mayo Foundation for Medical Education and Research.

Today the exterior, its high terraces and garden teahouse built of rough stone from the site, the mansion's cement-colored walls unpainted, has a distinct character of its own, more homelike than monumental despite its forty-two rooms.

Perched on a heavily wooded hillside, the house faces west, with occasional glimpses of the distant countryside, thanks to the high terrace on which it is built, though dense foliage often obscures a view of the mansion itself from the winding driveway, except in the lower section where the fields are stocked with wild animals such as deer and bison. Actually, it takes an aerial view to reveal the complexity of the structure with its multiple roofs—hipped, flat, and a combination of both—surmounted by a lookout tower and several dormers of unusual proportions.

From the exterior, the most notable features are the wide-spread eaves at several levels that shield from the sun the expansive rows of windows with both giant and tiny panes, the sudden breaking out of octagonal or half-round bays, and terraces high and low whose elevation varies in relation to the winding driveway that sweeps upward and around toward the rear of the mansion, passing at last the glass-paned doors to the garage built under the east end's balustered walkway before abruptly turning into the courtyard at the rear.

The unusual plans of the two principal floors combine ease of circulation with through ventilation and ample natural light—the last considerably improved in recent times by changes which Mrs. Charles W. Mayo had made, such as the installation of a large bay window which opened up the gallery to the courtyard and the replacement elsewhere of dead walls by glass blocks to bring light into shady corners without a sacrifice of privacy. On the main floor, the service units and dining rooms are separated from the private quarters and the adjoining foyer by the 48-foot living room and its parallel gallery which, back to back, both give onto the great outdoors.

It is the living room, of course, which is the heart of the house, containing, as it does, so much of beauty and interest that is revelatory of the outlook and aspirations, as well as the exquisite taste, of the people who for well over half a century have called it home. The 48 by 28-foot ceiling is heavily beamed in oak stained to match the paneling of the fireplace wall which houses ceiling-high bookcases at each end.

Opposite the fireplace is a huge, ceiling-high bay window—another of Mrs. Charles W. Mayo's projects—opening onto the main terrace whose Y-shaped flights of stone steps were the original approach from the driveway below. This window overlooks a branch of the Zumbro River and affords a panoramic view of the countryside beyond the towering trees. The

The gallery centered on a dining alcove.

Mayowood's main floor.

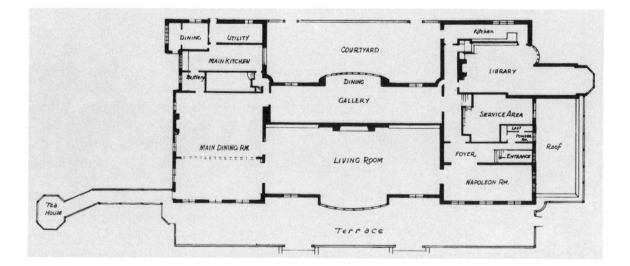

The main dining room minus
its extension.

principal fascination of the room, however, lies in its furnishings,
which combine antiques of various periods with turn-of-the-
century pieces, some wonderful silverware, and items gleaned
on journeys to remote parts of the world.

Most of the upholstered pieces are in sage green, except for
the wing chairs on either side of the fireplace which are covered
in a black cut velvet with a tan ground. The Chippendale side
chairs have petit-point seats, while the two Jacobean cane
backs are content with green cushions. In addition to the
fireside sofa and a grand piano, this room still provides space
for a collection of exotic pieces such as an eighteenth-century
Italian coffer, a credenza of the same vintage, a small Italian
chest, and a seventeenth-century Caroline gateleg table.

The long room called the gallery, which lies behind the
living room, actually performs two functions. As a wide hallway
it connects the two ends of the house, while its rear wall has a
French window at each end giving access to the rear courtyard.
Between these window-doors the gallery wall opens out into a
shallow area having a full-width rounded bay window overlooking the same statuary-bedecked area with its sculptured
central urn. This is the family dining room which has the same
red-tiled floor as the gallery so that there is no physical, or
even suggested division between the two. Indeed, the walls
between the gallery and the green toile curtains of the bow
window are a bare 4 feet long, though each one of them contains a hemispherical niche displaying a marble bust—one
of Apollo, the other of Athena. Below Apollo is a mahogany

62

OPPOSITE: A gay corner of the 48-foot living room.

console table on which reposes a Meissen tea set. The dining table is early nineteenth-century Hepplewhite, round-ended and extensible, of inlaid mahogany. In contrast, the chairs are of a Phyfe Empire design, painted white and covered in green and gold damask. At one end of this ensemble is an antique birdcage; at the other end is a small mahogany sideboard, so that there is very little room for navigation except in the gallery itself.

Off one end of this room, and next to the foyer, is a much smaller apartment called the Napoleon Room which contains several items of note, including mementos of the great French general after whom it was named by his admirer, Dr. Charles H. Mayo. This is quite a simple room which has been given character by an extraordinary Rococo door surround salvaged from a German castle bombed during World War II. Gilded, painted, and rubbed, it stands out against the pale green of the wall where light from the tall French windows of the main terrace can strike it.

In a corner between a green-silk window drapery and this bizarre doorway is an old-fashioned Aeolian organ whose pipes are attractively framed in an adjacent wall. Over the organ is a 1916 portrait of the founder of the Mayo medical dynasty, Dr. William W. Mayo, framed by silver candlesticks and a pair of colorful French porcelain urns. Next to these, on a cabinet built for music recordings, stands a mounted fan of Belgian lace reputed to have once belonged to Napoleon's wife, Josephine. Napoleon himself is celebrated by a large mezzotint, and in the opposite corner is a fine Sèvres vase on a tall stand.

Probably the finest piece of furniture in the Napoleon Room is a Louis XV *bureau plat* used as a desk, on which rests a French eighteenth-century bronze inkstand. France is also represented by a pair of Louis XV side chairs and one *fauteuil* in multicolored Aubusson tapestry. There is also a small canapé of the same period in striped blue-and-gold velvet, fronted by a table-style vitrine displaying small *objets d'art*.

Quite different from any of the rooms described thus far is the library, which is located on the other side of the foyer—a large, L-shaped room that is entirely paneled, including the principal beam and the mantelpiece, the latter being also decorated with chisel molding. These panels were fashioned from doors of the old school which all three doctors attended. Here, again, is a coved ceiling, all-white to reflect the light of a single chandelier, though ample daylight is captured by a huge bow window overlooking the side terrace.

On this floor are two large Turkish rugs of the Anatolian period in pink and gold, colors that, oddly enough, tone in splendidly with the upholstered pieces in green and their gold cushions. More pink is added by the marble top of a Louis XV *bombé* commode, lovely in its delicate marquetry and intricate ormolu mounts. Above it hangs a Mexican mirror in an ornate gold frame, with sentinels at either end in the form of exquisitely carved Philadelphia Chippendale side chairs.

The approach to
the summer house.

A family dining area off
the gallery.

OPPOSITE: In the master bedroom the two outer walls are
almost all window.

The Napoleon Room's portrait of the dynasty's founder and his desk.

Napoleon's portrait in the room named after him.

The rest of the furnishings emphasize the variety of furniture styles selected for their appeal without regard to period: the Regency card table banded in satinwood and the round-backed occasional chair cushioned in gray satin that goes with it; the Louis XV stool; a George III Canterbury; and an early eighteenth-century table with a pink-marble top—all seem quite at home with the Sheraton-type nest of tables, the modern, round-backed love seat all in green, and the ever-so-light pseudo-Queen Anne armchair with its looping arms and slender, curving legs, white-cushioned and gay.

Two other Mayowood rooms should be touched upon briefly as emphasizing variations in treatment that their functions make possible. One is the large dining room used for informal gatherings as well as large formal dinners. This seats up to fifty-two people and is furnished solely to that end, being adjacent to the main living room, and having casement windows opening onto the front terrace. This space actually consists of one long room which can be divided into two—one somewhat larger than the other—for smaller parties without a reshuffling of the furniture or having the guests pass through the closed-off section.

In both rooms the ceilings and trim are off-white, the walls a blue-green, with draperies of a blue and white print. On the floor is a large Savonnerie rug and two smaller Feraghans. The extensible dining table is English Chippendale, as are the chairs with their blue-velvet seats. The larger room, being at the inner end, is provided with a fireplace, on either side of which are large French windows, a single, larger window serving the other room section, and here it is that the furniture is normally set out. The principal pieces consist of a swell-front mahogany chest of drawers and a six-legged Hepplewhite sideboard, *circa* 1810, against the rear wall, which is adorned with a pair

of nineteenth-century French crystal sconces and two framed Chinese scrolls.

The other room is the master bedchamber which is over the Napoleon Room and part of the foyer so that it overlooks the garden on two sides, its corner location endowing it with outer walls that are almost all window. Though not overlarge it is light and airy. However, with walls an olive green, and the ceiling, trim, window pelmets, and a series of built-in closet doors plus two for the room all in white, the architectural character is somewhat severe because of contrasting rectangular areas and rigid lines. Such a setting calls for plenty of color and pattern, of folds and curves and contrasts both in shape and texture. And that is what these furnishings supply.

The rugs, which are a modern deep-pile type in gray with a fringe, call little attention to themselves and so form a neutral background for the pieces placed upon them. The dominating piece in this room is the bed—an English carved mahogany four-poster of 1850, with a white spread, lace tester, and dust ruffle.

The chest at the foot of the bed dates from 1774—the so-called Independence Era—and is painted in blue and gold, but the mahogany sewing table at the bedhead is mid-Victorian. Neither seems out of place alongside the pedestal table behind the sofa, whose twin leaves both drop so that it can be stored flat against a wall.

Such considerations of suitability, interest, and compatibility are characteristic of the whole house. Magnificence has been played down and a homelike atmosphere maintained over several generations of an active family that has not only made medical history but created and maintained an attractive home that still speaks of the high aspirations of its founder and those who followed after.

This exquisite Louis XV *bombé* commode stands in the library.

67

Hillsborough, California

HOUSE-ON-HILL

Eclectic Interiors Behind a Sixteenth-Century Façade

This Tudor mansion is most impressive from the rear.

The two-story-front's heavily planted forecourt.

The Tobin Clark mansion called House-on-Hill occupies a sunny 250 hillside acres with views of San Francisco's Lower Bay and the rolling landscape far to the west. Erected in 1931 on sharply sloping ground, this rambling, Tudor-style house is two stories high in front—if one includes the spacious rooms in the roof—and three at the rear, so that its principal rooms are of necessity on the second floor, an arrangement that is far from being the handicap that it may seem.

Since hills and winding roads are associated phenomena, the house is entirely hidden from view until one enters the stone-walled courtyard, where the mansion's U-shaped main façade occupies three sides of a walled-in forecourt four steps below the driveway level. On either side seemingly endless tile-roofed and dormered wings stretch out along the driveways, half hidden by the trees.

This central unit is of stone, its slate roof punctuated by two widely spaced dormers, with twin stone chimneys at each end. All windows have leaded casements, and the wide doorway with its bracketed segmental arch is crowned by a bust of the English poet Matthew Prior (1664–1721).

On either side of its wide brick path, the forecourt is heavily planted, and from it pleached and espaliered magnolias climb up to the finialed dormers, reaching toward one another along the lower roof and over the windows below.

From this point the driveway sweeps around to the back of the house which presents an entirely different aspect. In place of the Carmel stone of the front walls are decorative panels of mellowed brick filling the spaces between the oaken half-timbering, while the eaves are replaced by a deep cornice which supports three large cross gables between the wider, and taller, main gables of the H-shaped mansion.

Because of the sloping ground, the rear of the house is provided with a raised, flagged terrace enclosed by a low stone wall. In the middle of this wall, an opening between ball-topped pillars leads to five semicircular stone steps that become

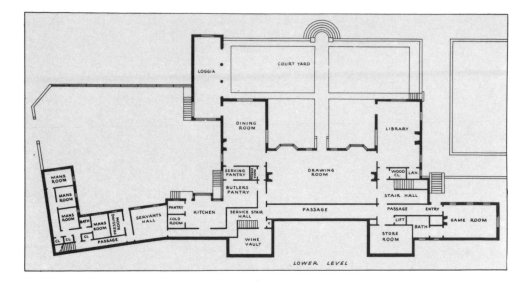

How the steep hillside location
affects the plan.

LOWER LEVEL

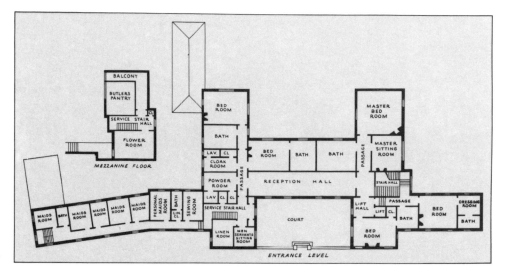

Odd-shaped floor tiles emphasize
the hall's perspective.

ENTRANCE LEVEL

A guest bedroom's striking
Chinese wallpaper.

A fascinating corner of
the sitting room.

progressivcly larger as they descend to the lawn where the land falls away in three directions.

Seen from the steps, the entrance porch seems cavernous with its overhanging bracketed bargeboards, the paneled double doors barely visible beneath the opening's carved architrave. The porch itself reaches back to the house wall, passing under a tiled pent roof that stretches across the central façade and, from a distance, looks like a framed overhang. Beyond these doors is the drawing room, for this is the lower level and over it is the front reception hall, which can be reached by a wide stairway or even an elevator.

A much more welcoming entrance to the 55-foot-long reception hall is via the front door which opens directly into it. This is one of the most striking interiors in the house, thanks largely to the transverse ceiling beams, which, however, do little to shorten it visually, and the odd-shaped, black-and-white floor tiles which emphasize the perspective more than any Hungarian-point parquet ever could.

The walls are of particular interest, consisting of antique, Jacobean quarter-sawed oak panels, punctuated by fluted Ionic pilasters on paneled plinths, and reaching to within frieze height of the ceiling. From the two end ceiling panels hang modern six-branched chandeliers of pewter with ball finials having a sturdy Cromwellian appearance, augmented by four wall fixtures with hurricane globes and glass smoke bells. The furnishings include a pair of Chinese lacquered cabinets in ebony, with pictorial drawer fronts in gold and silver. These are mounted on extravagantly carved, Baroque-style, English stands. Opposite them are a pair of Georgian black-and-white-veined marble-topped console tables, supported by black, bronze spread eagles mounted on black-marble plinths. Between the Chinese cabinets is a transitional Louis XV–XVI *bureau plat*. The set of eight Queen Anne design high-backed side chairs of pine, with cabriole legs and *pieds de biche*, embellished in front with a cartouche-shaped scroll are fitted with loose cushions in cardinal-red velvet.

On the lower floor is the library, a beautiful room with a 15-foot ceiling and walls of antique pine, with a paneled dado and deep, coved cornice. The fireplace wall, which projects beyond the room-high bookcases, has an overmantel flanked by Gibbons-style carved drops, the fireplace itself being enframed by a wide bolection molding in *verde antico*, surmounted by a triple-section Venetian mirror in gold leaf.

The floor is laid in a simple parquetry pattern, largely covered by a modern carpet in a medium brown which forms an excellent background for the upholstered pieces—sofa, open and closed armchairs, and ottomans, all in a pale green with a beige fringe, along with a pair of old-time wing chairs elegantly styled in brass-studded brown leather.

Apart from the five-piece fireplace seating group there are plenty of other antiques to give the room an early, if not quite Tudor, atmosphere. These include a colorful eighteenth-

One aspect of the huge
drawing room with its walls
of bleached pine.

century English pole-type fire screen in red, blue, brown, and beige needlework and a long mahogany Chippendale library table with a terrestrial globe on one end and a celestial one on the other.

Closer to the Tudor period is a pair of Cromwellian side tables with turned legs and stretchers, but equally interesting are the early eighteenth-century Queen Anne miniature four-drawer bureau centered on the library table; a drop-leaf, spider-legged table, which stands behind the sofa; a Queen Anne side chair in green leather, with a nice English breakfront bookcase, and, best of all, a sixteenth-century gateleg table in oak.

Next door is the huge drawing room (27 by 55 feet) which has a light-tan ceiling decorated all over with stucco moldings, a treatment so often necessary both for architectural balance and decorative coherence, as well as to break up the vast areas and so reduce their apparent dimensions.

The floor is parquet of English origin, covered here and there by a series of gorgeous Oriental rugs, disposed at wide intervals lengthwise with the room. This room has a fireplace at each end, set in white-marble bolection moldings, with black and white hearths.

At each end of one longitudinal wall is a fabulous Coromandel screen in colors ranging from gold to crimson, including a green comparable with that of the three large sofas, and closely imitated in the velvet upholstery of a *voyeuse* signed by the great French *menuisier* Georges Jacob (1739–1814). Alongside each

71

The dining room with its eighteenth-century paper and lovely mantel.

Each bedroom has its fireplace, bath, and dressing room.

screen is a voluptuous carved and gilded console with a rectangular black-marble top—each one of a different style characteristic of English work. Other touches of color are provided by a pair of Louis XVI *fauteuils* in tapestry and two brass-studded pairs of the Louis XV style in needlepoint. Still a different accent is introduced by a Chinese coffee table in ebony, as well as by a pair of wing chairs in a two-color, stencil-patterned silk, their high backs rolled in the same manner as the arms, suggestive of a William and Mary extravaganza.

Next in importance is the dining room which goes to the other extreme as regards furnishing, though the decoration is superb. In this room the ceiling is off-white and plain; the trim also is white, with an elegant modillioned cornice and deep, molded baseboard, plus a forty-five-degree panel filling each corner of the interior. The walls, from baseboard to frieze are covered with a lovely eighteenth-century Chinese wallpaper having an overall pattern of climbing willows in full bloom, among which birds of all colors and sizes, from peacocks to swallows and birds of paradise, are disporting themselves.

The mahogany dining table is a fine example of late seventeenth-century Hepplewhite in a pedestal style, with a smaller circular one by the window embrasure which is used for informal meals. The serpentine-front Sheraton sideboard, with its high brass rail, is in three sections and has twelve square, tapered legs.

A Chinese wallpaper similar to that of the dining room is used in one of the guest bedrooms, but here the predominant colors are varying blues on a white ground, with touches of red and yellow. The rest of the room, however, has the early American Georgian features such as the key-molded cornice, raised panels in the walls, doors, overdoors, and dado, all painted white, with overall carpeting in gray.

Quite naturally, most of the room space is taken up by a pair of beds. In this instance their head- and footboards are covered in a blue and gray silk damask, the same material being used to cover a low-backed armchair in the Chippendale style.

Some of the other bedrooms in this house are far more modern, such is the adaptability of these pseudo-Tudor interiors. Probably the most extreme example is the master bedroom, which has overstuffed furniture and, spread over the beautiful parquetry floor, a beige sculptured rug in a giant stepped pattern.

Such need not be said of the high bed's decorated posts which, parallel-sided as they are, look more like the twin poles used to uphold a tent awning on a medieval battlefield. In any event they do interject a twelfth-century note into the room, their tassel-like finials suggestive of ceremonial spearheads, while their decoration in gold and white, together with the arching framework of a nonexistent canopy somehow seems far more at home than the modern pieces ever could be with the exotic wall covering, whose beauty of line and color is suggestive of a Louis XVI silk hanging.

OPPOSITE: The library with its paneled walls of antique pine is comfortable, yet stately.

Waltham, Massachusetts

GORE PLACE

A Federal in New England

Among the more notable American houses built since the turn of the nineteenth century, Gore Place has the distinction of combining a remarkable history with a novel plan and a decidedly imposing appearance, in spite of a lack of specifically decorative features either inside or out.

The story of its origin goes back to 1799 when Christopher Gore and his wife, Rebecca, were in London where he had served as a commissioner under the Jay Treaty and chargé d'affaires. At this time his original country house—a wooden structure—was destroyed by fire, and immediately thereafter Rebecca Gore, displaying an unusual talent for such things, began planning the much more substantial house, for eventual use as a permanent residence, that was to replace it. In 1801, when she and her husband were in Paris, she consulted a noted French architect, J. G. Legrand, who translated her rough sketches of the desired accommodations into working drawings.

In 1805 construction was begun on the new mansion which was to incorporate ideas adapted from certain English and French country houses, including an oval dining room projecting half its depth beyond the rear façade and a state reception hall, centered in the entrance front, whose inner wall was to be semielliptical in plan. In the finished house these two large rooms with extra-high ceilings occupy a major part of the two-and-a-half-story central block from each end of which extends a one-and-a-half-story wing terminating in a pedimented pavilion.

One of the most interesting features of these façades is the size and placement of the doors and windows, some of which seem to bear little relation to the interior floors. In the north, or front, entrance, for example, there are two doors on a level with the wide stone terrace, one at each end of this flat façade. One of these doors gives access to a square room known as the Butler's Hall, the other to a room of similar size largely occupied by the circular flying staircase.

Between these two wide doorways, whose already large

Imposing Gore Place
on its hilltop.

The north façade.

OPPOSITE TOP: A feature of the music room is an exquisite *trumeau* setting off the Adam-style mantel.

OPPOSITE BOTTOM: A study designed for visual pleasure and personal comfort.

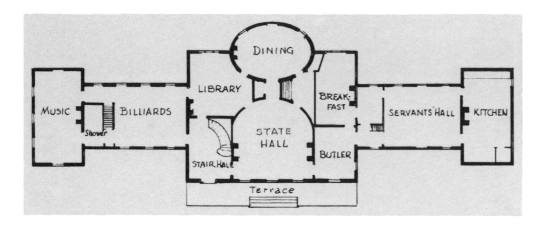

The plan centered on an oval dining room and D-shaped state hall.

Star-inlaid handsome kneehole chest of drawers, *circa* 1740, with matching dressing stand.

The tiled stair hall with its beautiful eighteenth-century English lantern.

dimensions are emphasized by a half-round fanlight spanning both the door and its sidelights, are spaced three tall double windows, quite narrow in proportion largely because of their slender trim. Over each of these five openings is a normal sash window, tucked close under the overhanging eaves, whose sill is half the height of a room above the windows below. The arrangement exposes vast areas of mellow pink brickwork of a particularly pleasing texture thanks to the bricks being set in Flemish bond—alternate headers and stretchers—which eliminates any variation in the pattern.

All of the window and door trim is of sand-colored wood, set within the reveals except for the tall hooded windows of the center section. On the low-pitched hipped roof above, a pair of square-faced shed dormers extends back to the two twin chimneys, while the roof peak is crowned by a large but low octagonal cupola which admits light to the stair well.

In contrast with this arrangement, all the windows in the wings are set into tall arched recesses, the rectangular sash being narrower than the brick setbacks whose arches are fully occupied by the half-round second-floor windows. These fit snugly under the eaves and serve to light the upper rooms which extend partway into the roof. The single-room pavilions have rectangular recesses in the gables, but only one is pierced by the first- and second-floor windows. In each case there is a tall chimney at the junction of the wing and pavilion roofs which serves a fireplace in the latter.

On the south, or rear, façade, the distribution of these voids is quite similar, but the total aspect is changed, somewhat for the better, by the projecting rounded wall of the oval dining room. Apart from this, the main-block window arrangements are quite similar except for the fact that the twin entrances consist of French windows in place of paneled doors, and all the rectangular windows are fitted with louvered blinds. Across the principal façade of the main block extends a deep sandstone terrace of sufficient height to accommodate both horsemen and coaches, with a flight of steps in the middle leading down to the driveway.

The total result of this carefully worked-out design is an imposing house of twenty-two rooms, with a façade 175 feet long that looks much larger than it is and is quite unlike that of any other great New England residence of its period. In the beginning, the grounds, too, were unusual in that they were laid out in the manner advocated by Humphrey Repton, the leading English landscape architect of the period, utilizing a great variety of native American trees and shrubs, cultivating fruit orchards, and planting hawthorn-hedged walks, with little attention paid to the rearing of flowering plants, except for specified areas reserved for cutting garden and herb garden.

Today the estate's 400 acres have shrunk to the original 40-acre homestead lot; its dependencies have been reduced to a single coach house and farmer's cottage. Nevertheless, in the hands of the present owners, the Gore Place Society, the

The state reception hall
is impressive in its dignity.

Over the billiard room
is this twin-roomed nursery.

The entrance hall's antique mahogany tall clock.

The secretary's dainty room.

property has been restored to its former grandeur. The rooms have been refurbished, and many of the original furniture pieces, as well as books and portraits, have been retrieved so that the interiors recapture the atmosphere of those years from 1809 through 1817, when, after serving as governor of Massachusetts in 1809 and 1810, and as United States Senator from 1813 to 1816, Christopher Gore retired from public life to end his days as a simple country gentleman on his Waltham estate overlooking the Charles River.

The principal entrance to the house is the wide doorway at the left-hand end of the terrace. This opens into a large stair hall paved with random black-and-white marble squares. Despite its unusual height, this room is entirely devoid of frieze or cornice, the only decorative features being the trim of the round-topped doorway with its slatted fanlight of leaded glass. On the other hand the staircase is extremely beautiful in its simplicity. The steps and plain square balusters are painted white, with a simple mahogany handrail ascending in an exquisite curve, in line with a dado molding attached to the rounded wall. A splendid focal point is provided by a delicate, brass, eighteenth-century English lantern, centered in the stairwell, which illuminates the whole area.

Furniture here is reduced to a minimum, in keeping with the period—a mahogany tall clock by Joachim Hill (1810–1850) of New Jersey, a 1790 New England sideboard table with satinwood panels displaying a pair of Oriental blue-and-white porcelain ginger jars (*circa* 1722), under a carved and gilded eighteenth-century Chippendale looking glass and, flanking the entrance door, a pair of colorful Chinese garden seats.

The area opens into the billiard room, the library, and the State Hall which acts as a reception room and has the same marble-tiled floor as the entrance hall. This floor is entirely devoid of rugs, preserving the air of austerity created by the plain walls and ceiling and floor-length tall windows, emphasized somewhat by the full-length draperies of yellow and ivory lampas in a cornucopia design.

At each end of this room is a fireplace, one connected to the other by a deeply curving rear wall which is pierced by doorways giving access to the breakfast room, oval dining room, and library. In the center of the floor is a large, round mahogany gateleg table of English provenance, *circa* 1800, which belonged to Governor Gore, and on it is a lovely early-nineteenth-century Chinese Export bowl in the sacred bird and lotus pattern. The chairs are eighteenth-century Chinese Chippendale pieces covered in orange silk.

Back to back with this room is the oval state dining room, with curved doors and mantel and the same high ceiling, but papered in a warm gray with alternating tiny medallions and dots in white above a white-painted dado. This inspired wall covering makes the room seem even larger than it is, the tall windows adding to the feel of spaciousness. Hanging in straight folds alongside them are draperies of self-striped blue silk with a blue fringe of a shade reflected in the great Oriental rug patterned with flowers in rose tints on a blue ground.

From the center of the ceiling hangs a basket-pattern chandelier of a Louis XVI design from a church in Tours, and under it is an accordion-type dining table—a Philadelphia Sheraton design of 1810. On this table are grouped a Sheffield silver plateau and epergne whose cut-glass bowls brim over with grapes, plus a pair of twin-taper silver candelabra. At the end opposite the fireplace is the Gore family's Boston-made Hepplewhite sideboard, of the swell-front type, with a bird's-eye-maple inlay and beautifully grained mahogany doors. On it rest a pair of English slant-top knife boxes with a silver tureen in the middle. On the wall behind it is an English gilt girandole with twin sconces and a bellflower ornament below to balance the tall eagle cresting.

The breakfast room is a much smaller rectangular apartment and far less formal, the wood trim white against putty-colored walls, with a Tabriz rug in tones of beige and blue, and hangings of a cotton floral print in muted tones on a putty ground. In this setting, Governor Gore's own English twin-pedestal oval breakfast table (*circa* 1800) looks its lovely best, attended by English Hepplewhite-style inlaid mahogany chairs in the wheat-sheaf design of 1790, with brown leather seats and arm pads. Here also is an authentic Sheraton sideboard of inlaid mahogany, dated 1800.

The Boston-made Hepplewhite sideboard.

A "concertina" table, fully extended, accommodates a long silver plateau beneath the basket-pattern chandelier.

A Hepplewhite-style bookcase-secretary by Lebbeus Bailey of Maine.

In the window is another Gore piece, a pedestal-type, four-footed mahogany tea table (1800–1805) that serves to hold the tea caddy and dessert epergne, while a 1760 mahogany lowboy displays a teapot—bearing the Gore coat of arms and marked "T. Robins," a London piece of 1803—and a lacquered tea tray.

Quite different in style is the music room at the end of the left wing, though it exhibits less architectural detail and the fanlight is spaced well above the window. This room, too, has off-white trim and walls, but the window draperies are of a rich red silk with a woven border, and the rug is an Isfahan in varied blues and reds.

Actually, the room is large enough to boast of twin Waterford-crystal and ormolu chandeliers, supposedly from the Imperial Palace at Pavlovsk, which add considerably to its character. The fireplace also is impressive, being of white-painted wood, carved in the Adam manner around a sandstone facing. The projecting chimney breast also finds room for a *trumeau*, consisting of a large painting of an English landscape over a long horizontal mirror, all in a decorative gilt frame. The only musical instrument is an English pianoforte, *circa* 1800, in inlaid mahogany veneer, which occupies a corner between the front and side windows.

One particularly attractive grouping consists of a large Sheraton sofa in a blue-and-ivory-striped fabric, in front of

The informal breakfast room with the governor's own table.

which there is a lovely Queen Anne tea table (1730–1735) with silver sticks on the slides and a pair of painted English Sheraton chairs with cane seats, *circa* 1800, which belonged to the Gores. Over the sofa hangs a portrait of Mrs. Joseph Musgrave, which Sir Joshua Reynolds painted in 1752, flanked by three silver sconces. Another group takes in the hearth with its mahogany-framed fire screen covered in silver, blue, and gold antique fabric, a Martha Washington chair in gold silk brocade, and a wing chair of 1780 in red cotton brocade which almost hides an inlaid lyre-based corner table of Baltimore origin. Other eye-catchers are the tiny triple looking glasses in gilt *chinoiserie* frames on each side of the mantel.

As a lesson in contrast, the governor's study has much to offer, not only as a room devoted to work but also as an interior designed both for visual pleasure and personal comfort. This oval room is directly over the dining room and is lighted by three large, deeply recessed windows, with bottoms paneled in white to match the dado, carved mantelpiece, and ceiling. Over the eggshell dimity curtains are crimson ribbed-silk hangings with a black and gold fringe and tassels. These are hung in a deep swag with sill-length cascades that reveal the gold-tinted lining. The ceiling-high wallpaper displays gray and white rosettes in a diaper pattern on a putty ground— a perfect foil for the Serabend rug with its rose-colored overall pattern on a ground of pale beige.

This color scheme is extended to Empire-style chairs in red silk with gold medallions. The Hepplewhite sofa is upholstered in pale-blue, ivory, and red stripes, while the Queen Anne daybed displays crewelwork embroidered in blue and rose on an ivory background.

Turning to the more practical aspects of the study, the governor's paper work was undoubtedly done on the large drop-leaf table behind the sofa, on which a portable writing desk in gilt-decorated black lacquer now rests. This table, however, is but the central section of an extensible dining table once owned by the governor's brother, Samuel. Its removable, semielliptical ends now serve as consoles against the bowed wall between the windows.

Finally, at the far end of the room is its most imposing piece, a Hepplewhite-style secretary-bookcase, made around 1800, incorporating a "kidney-dial" clock by Lebbeus Bailey of North Yarmouth, Maine, and containing some of Christopher Gore's precious books which he once wrote of as constituting "a comfortable resource of age."

Features of the library are this bookcase and a gaming table.

Mrs. Gore's top-floor bedchamber.

Houston, Texas
BAYOU BEND

A Latin Classic With Period Rooms

Shadows dapple the garden
façade in the morning sun.

Anyone seeing Bayou Bend for the first time will recognize that this splendid residence is something out of the ordinary both as regards its architecture and its room arrangement, blending to perfection, as it does, the Palladian country-house plan with the best of Mediterranean and American Colonial features as adapted to the subtropical climate of the Gulf Coast. And as regards its interior design, the story is the same.

In eclectic furnishing the usual practice is to mix periods and styles in every room, and with a certain amount of knowledge, combined with good judgment, such a system can be most successful in creating interiors that are not only attractive and livable but which also conform to any desired level of formality. On the other hand, there is an interesting alternative, as demonstrated in this notable Texan mansion, where each of the nineteen most important rooms is representative of some particular period or style of furnishing, ranging from the late seventeenth century to the mid-eighteen hundreds.

The house, therefore, is totally different from anything one is likely to encounter, casually or otherwise, a fact that can only be explained by its most unusual history—and the character and interests of its original owner who had it built, with an exciting plan in mind, as recently as the late nineteen-twenties. For one thing, its façades reveal a Spanish influence (a style so well adapted to the locality), on a basically Palladian plan consisting of a porticoed central block with flankers connected to it by single-story hyphens, and the whole exterior faced with stucco given a pinkish tinge by the addition of crushed rose quartz.

Although built in the twentieth century, the whole structure has a faint Greek Revival air thanks to the parapeted walls and the monumental chimneys centered over the front walls of the flankers, plus the Greek ornaments of the central portico with its four two-story, parallel-sided pillars, innocent of capitals or bases and therefore more Southern plantation in style than purely classical. Nevertheless, the whole principal façade, with

OPPOSITE: The seventeenth-century Pine Room furnished in William-and-Mary style.

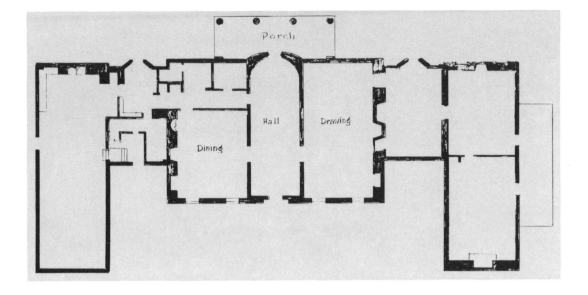

The house's open floor plan
is perfect for the display
of its treasures.

its nicely proportioned masses, its square-topped windows and doors, peering through a screen of tall trees and foundation bushes, is captivating in the extreme and rendered even more imposing in the sunlight by the copper-roofed hyphens.

The main entrance doorway is particularly effective with its heavily paneled door, surmounted by a bull's-eye fanlight and enframed between tapering, and slightly crossetted, stiles. There is a vertically grooved overdoor which is topped by a delicate scroll molding with an anthemion ornament at its center, a decorative detail repeated over the French doors centered in the hyphens. Above this door is a nicely designed wrought-iron balcony of New Orleans origin, accessible through a second-floor French window.

Standing in fourteen acres of gardens, the house is set off by a sloping front lawn down whose center sweeps a wide flight of shallow steps marked out by flowering borders punctuated with tall urns and stone benches. Around this lawn's perimeter are clipped hedges, creeper-smothered brick pillars and walls that, along with the mansion's terrace plantings of azaleas and camellias, invest the whole area with a feeling of antiquity, against a background of moss-draped live oaks, tall pines, and deeply shadowed evergreens, all strongly three-dimensional in the bright sun. At the rear is a woodland garden with natural paths, camellias, azaleas, dogwood, redbud, and a myriad of daffodils and narcissuses, completing a perfect setting for the house and the treasures it contains.

Such is the entrancing U-shaped house designed for Miss Ima Hogg, daughter of the famed Texas governor of 1891–1895 James S. Hogg, by architect John Staub, who supervised its erection in 1927 and 1928 and helped design the room backgrounds in keeping with the antiquities, decidedly mixed as to period, with which they were to be furnished. An avid collector of antiques since 1920, Miss Hogg, in making this house her home, expended much time and effort during the next thirty-odd years in acquiring the finest of American period pieces and accessories dating anywhere from 1680 to the Victorian era

An entrance hall as beautiful and homelike
as any room.

OPPOSITE: Dufour and Le Roi wallpapers emphasize
the music room's Federal air.

The Murphy Room displays seventeenth-century styles on an early eighteenth-century painted floor.

Opposite the Murphy Room fireplace is this ancient court cupboard.

which allowed for the inclusion of many Texan heirlooms.

Since it had long been Miss Hogg's intention eventually to make the collections accessible to the public, she began, in 1958, to decorate the rooms in a manner historically correct for articles of the same general style and period with which they were to be furnished. This involved some modifications to the interior architecture and finishes of some of the rooms in order to provide the furnishings with authentic backgrounds of the period represented. This work was completed by 1965 when she turned over the entire estate—house, grounds, and furnishings—to the Museum of Fine Arts of Houston by whom it is now administered.

Bayou Bend is today therefore one great house which, though entirely livable, is free of the clutter attendant upon family occupation, preserving in their entirety nineteen different styles of interiors that well-to-do citizens of the country's principal geographical areas were wont to create for themselves during certain progressive periods in the nation's history, with particular accent upon American furniture, although there are some notable exceptions, especially in the fabrics and ceramics.

The earliest of the Bayou Bend interiors is that of the Murphy Room, named after the late Katherine Prentis Murphy, noted antiquarian of New England. Since the furnishings of this interior are a combination of William and Mary (1650–1702) and somewhat earlier seventeenth-century pieces, the fireplace wall was paneled and painted an old-time red matching that of a young man's coat in a portrait group of the Hudson Valley school which hangs over a 1690 Connecticut "sunflower" chest. The wall paneling incorporates two shelved recesses with cupboards below which are used for the display of delftware and pewter. In contrast, the rest of the walls are whitewashed between a heavy molded cornice and a plain baseboard, both of which are finished in a reddish brown.

The ceiling, entirely innocent of exposed timbers, as one would expect in a house of this quality, is of white plaster, and the oak floor is painted in large black and white squares set diagonally (in quincunx style) to create an impression of greater size. The two large windows are draped with floor-length curtains in an eighteenth-century blue-on-white crewel work, with pelmets of the same material.

The furniture pieces themselves range from a Pilgrim slat-back mushroom chair and a light, turned, Carver armchair, both with rush seats, to Spanish-footed banister-backs of 1700–1710 and a roll-arm chair with Spanish feet of the same vintage. Under one of the windows stands a distinctly unusual table with widely splayed turned legs, having long-toed Spanish feet to match and plain stretchers. Here also is a Spanish-footed circular drop-leaf table, a small oval occasional table, and a turned-base tavern table with a breadboard top in cherry. Relegated to a corner by the fireplace, an early eighteenth-century daybed displays a padded head, while at the opposite end of the room stands a magnificent Connecticut-made court cupboard decorated with tulip carvings, split turnings, and small turtlebacks, or bosses, *circa* 1680.

Every bit as colorful as this interior is the Pine Room, furnished for informal meals. This room has considerably more panache and, being completely paneled, may even be slightly earlier. Actually, the raised paneling is 1927 vintage, copied from that of a Rhode Island house. Nevertheless, in its natural pine coloring it forms a perfect background for the pine, oak, and burl-walnut furniture, most of which is in the William and Mary style.

The fireplace is of special interest, being faced with blue-and-white antique delft tiles, held in place by a wide bolection molding. The overmantel consists of twin, extra-wide panels flanked by fluted pilasters, the caps and bases *en ressaut*, and the cornice a simple ogee molding. Centered in the off-white ceiling is a lovely English chandelier of brass with an eagle-ornamented stem, dating from the 1750's.

The floor is of random pine covered by a huge Feraghan rug in blue and red on a beige ground, the red being closely duplicated by the rusty brown of the velvet draperies which, together with the deep, scalloped pelmets, are edged in red-dotted white braid. Outstanding in this rich setting is a William and Mary six-legged tallboy in handsome burl-walnut veneer, complete with cup-turned legs and scrolled stretchers, made even more eye-catching by the yellow English slipware hanging over it.

The same burl-walnut veneer finish is displayed by a bun-footed slant-top desk with silver accessories in a window-lighted corner. The oval drop-leaf table, however, is of maple

The Federal parlor and its Adam mantel.

A corner of the Queen Anne sitting room.

In the Newport Room are displayed
masterpieces of the Townsend-Goddard school.

with beautiful turnings and unusually long square sections in the legs. At either end of it are Spanish-footed, cane-back armchairs, one with a Flemish cresting and ram's-horn arms, the other having a concave crest. Both have turned stretchers and thick pad cushions in pink wool damask over their cane seats. Other chairs include a roundabout with pierced slats, turned legs and stretchers, and rush seat; a rush-seated, Spanish-footed banister-back armchair; and an early-eighteenth-century wing chair in cream-colored leather, brass-nailed. Of exceptional interest is a pair of mezzotint portraits of King William and Queen Mary which hang over the fireplace.

In the Queen Anne sitting room upstairs, the quality of the furniture becomes increasingly elegant, though on the average only slightly more recent, because of the fact that the Queen Anne-style furniture was made either in Boston or Philadelphia, from earlier English designs, toward the beginning of the eighteenth century. By mid-century, japanned maple or walnut pieces imitating Oriental decorative techniques had become quite popular. The Queen Anne chairs in this room were probably made about 1740 and then were sent to China about 1800 to be "modernized" with paint decoration. The fabulous highboy, *circa* 1745–1750, is also decorated in this fashion, and is supposed to be of Boston provenance. The plainer Queen Anne chairs, including one armchair and a daybed, were made in Philadelphia at about the same time, as was a corner chair with its original cowhide covering and slipper feet and a small but handsome trifid-footed desk-on-frame. All the rest of the pieces are pad-footed, except a tea table from Newport, whose slipper feet and slender legs give it quite a saucy air but in no way detract from the gilded black Astbury-ware with which it is laden.

The several other furniture items in this room, which add to its informal air, consist of a Philadelphia wing chair, something of a rarity because of its rear cabriole legs, and a handsome inlaid card table which incorporates an accordion action enabling it to be extended at will. The setting in which these pieces are arranged is made up of white walls, with the doors,

One of the twin, carved shell
cupboards in the Newport Room.

OPPOSITE: The mid-Victorian Texas Room is wholly lined
in cedar.

89

In the Queen Anne sitting room is this japanned group, *circa* 1780.

A corner of the Queen Anne sitting room.

windows, and dado of pine, plus a paneled fireplace wall, the mantel itself having an eared frame around the facing of delft tiles, under a pillow-block molding dentiled below the shelf.

Next in the scale of progression is the Newport Room which displays some of the finest work of the Townsend-Goddard school. Its paneled fireplace wall and dado were reproduced from Library of Congress drawings of the Hunter House in Newport and include a pair of richly carved shell cupboards now full of seventeenth- and eighteenth-century American silverware. An unusual and most interesting feature of the cupboards is the insertion of the head and wings of a cherub, carved in the round, into each of the spandrels of the cupboard arches. These heads are painted in natural colors, while the rest of the woodwork is finished in apple green. The fireplace itself is flanked by fluted Corinthian pilasters, and the cornices are deep cavetto moldings from which the heads of the pilasters and cupboard arch keystones project *en ressaut*.

At the other side of the fireplace is a cabriole-legged tea table which has a dished top and normally accommodates a polychrome salt-glaze tea set. Nearby is a mahogany lowboy whose plainness is relieved by lovely recessed shell carvings and notched corners to the top. Ball and claw feet much less crisp in the carving are a feature of the Martha Washington-type chair which faces, across the fireplace, a Chippendale wing chair with Marlborough legs and vertically rolled arms; one chair is in yellow damask, the other in a salmon-pink brocade adding cheerful grace notes to a rich interior.

Probably the most elegant of all these early interiors is that known as the Massachusetts Room, whose wholly paneled walls are painted a deep blue of a shade inspired by an eighteenth-century Portuguese chintz. The fireplace itself is a study in design, its nicely proportioned horizontal, double-beaded panels set between fluted pilasters on wall pedestals slightly above windowsill height. The fireplace opening is enlivened by a facing of Fazackerly-type Liverpool tiles depicting birds and flowers on a white ground.

In this room the furniture is Chippendale-style, made in Massachusetts, and centered on a seventeenth-century Caucasian rug of a highly stylized dragon pattern in blues, reds, and yellows—colors picked up by the heavy, fringed brocatelle draperies which have fretted valances of a palmette design with a distinctly Oriental cast. On either side of the fireplace is a cabriole-legged wing chair, one of them covered with its original flame-stitch fabric, the other in a reproduction silk brocade. The side chairs are a set, all having the same Dutch underbracing, pierced splat, shell crest, and carved knees as the double-backed settee which stands beneath a portrait of Mrs. Winslow and Child by Joseph Blackburn. In front of this settee is a particularly interesting Massachusetts-made tea table with the rare turreted frame which gives the top its scalloped shape. Elsewhere, serving as side tables, are two English dumbwaiters, one with revolving shelves and a four-legged pedestal, the other the more usual tripod affair with ball-and-claw feet and three stages. These and the tea table display examples of Wedgwood and Whieldon china, but two other pedestal tables of the birdcage type with piecrust tops are used as flower stands. This is one kind of room that survived the American Revolution despite the English origins of the furniture designs which were to remain popular for centuries to come as exemplars of good taste, even though displaced temporarily, in some instances, by the somewhat more sophisticated Federal styles with their republican motifs.

Another, and slightly later, example of the same period that is even more stylish is represented by the drawing room. Here, as elsewhere, it is the architectural elements that determine the setting and give a clue to the period represented. In this case the interior was patterned after the Neo-Palladian rooms then coming into fashion, when some of the most beautiful and ornate Chippendale designs were being produced. The walls of this room, therefore, were not paneled but given a substantial molded baseboard and a wide, decorative chair rail over the plaster walls, both painted a golden yellow. The cornice was well developed, with dentils and egg-and-dart molding below the supporting modillions. The floor-length window trim was neat and unobtrusive, slightly crossetted, but doorways were surmounted by a pillow block below the wide-spreading molded cornice, as in the case of that great Southern mansion, Shirley,

A stoneware and Wedgwood creamware display.

The dining room with its painted wall covering.

The eighteenth-century
Massachusetts Room.

A Peale landscape is featured in
the McIntire Room.

which was built in 1769. Formality was therefore the order of
the day, as a glance at the furnishings will indicate.

This room deliberately mixes Queen Anne and Chippendale
pieces, both made by American craftsmen, set off by a nineteenth-century Kirman rug in beige, rose, and blue and draperies of gold damask, contrasting slightly with the gold-tinted
walls. There is also a pair of uncomplicated globe-type English
chandeliers of 1780 with star-shaped *bobêches* and no more than
a dozen crystal drops to the branch. Dominating the room is a
beautiful Newport bonnet-top secretary which finds little
competition with an adjacent tall clock with the same type of
finials. One notable variation is introduced by the block-
footed Philadelphia Chippendale sofa in silk damask—the
centerpieces in a balanced wall arrangement which includes a
pair of cabriole-legged card tables below an exquisite pair of
Chippendale looking glasses, on either side of a family portrait
by the great John Hesselius. All the chair backs, incidentally,
are of a different design, though all have cabriole legs except
one block-footed armchair which seems to combine Chinese
design with Gothic touches plus a beaded top rail.

From this point on, the rooms emphasize the changing
fashions of the Federal period which was to embrace so much
of both Adam and Phyfe. One of these is the music room in
which Dufour and LeRoi wallpapers of 1828, combined with
pale-blue walls, have a notable impact. These walls are
enframed by a deep white cornice and white dado, the latter
including a chair-rail molding and deep skirting board. On the
fireplace wall is a white Federal mantel incorporating the
American eagle-and-dove-patterned Etruscan lamps, around a
facing of mottled yellow marble with a pierced brass fire basket
inside it. Over this is a Federal-type mirror with a convex
sunburst center, flanked by a pair of eagle-motif, twin-candle

sconces which leave most of the artificial lighting to a vase-type eighteenth-century crystal chandelier. At the windows are fringed cream-silk curtains, overhung with cascading pelmets overlapping at the center. Picking up the wallpaper colors is a magnificent Bijar carpet with a floral design in cream and old rose on deep blue.

The major furniture pieces consist of McIntire-style side chairs with light-mulberry-tinted seats; a lovely gaming table with slender, turned legs and ivory pulls, attributed to William Hook; a fine Phyfe piano; and an early-nineteenth-century wing chair in mulberry-ribbed silk, all contributing to a well-nigh faultless period interior at the end of the nineteenth century's first quarter.

Elaborating on the Federal theme are the dining room with its Phyfe pieces and Salem Hepplewhite adaptations; the McIntire Room focused on New England; and the Federal parlor with its Adam-style mantel and pastel Aubusson rug, culminating in the Mid-Victorian Texas Room, cedar-boarded all over (including the ceiling) with Gothic arches. Currier and Ives prints decorate the walls, and a portrait of General Hogg, Miss Hogg's grandfather, hangs over a country side-board from Tennessee. This is loaded with a fine display of late Staffordshire "Texian Campaigne" ware that also fills the wall recesses and covers the dining-room table in this most modern of Bayou Bend interiors which shed considerable light on the changes in furniture and interior design in the higher social levels from Late Colonial to Mid-Victorian times.

Among the furnishings is this pair of 1795 Salem shield-back chairs by McIntire.

The drawing room with its family portrait by Hesselius.

Denver, Colorado

THE MINISSALE RESIDENCE

An American Tudor Palazzo

The mansion from the water garden.

Looking for all the world like a sixteenth-century English Tudor manor house attached to a half-timbered annex, the Joseph N. Minissale residence in Denver is partially screened by limestone additions bearing the proud hallmark of Italian workmanship in its exquisite carving and sculptured detail—a combination that is almost without parallel in the history of architecture. Another outstanding feature of this attractive cluster of buildings is the total absence of balance or symmetry, either horizontal or vertical, so that the viewer is constantly discovering new viewpoints and fresh conceits to admire while seeking in vain to pinpoint the unifying factor that invests the whole structure with interest if not enchantment.

Obviously Elizabethan are such details as the spindling chimneys, triple and quadruple, twisted, fluted, hatched, and almost always crowned with horizontal bristling spikes; the steep roofs of mottled slate; the walls of brick with burned headers set in diaper pattern and capped with tooled masonry; and the mullioned windows overhung with carved dripstones. Here, however, are windows of several kinds—flat-topped, round-topped, and cusped, some set in an oriel, others in a semioctagonal attached and battlemented tower that climbs a chimneyed gable wall. Only the wide entrance doorway has a Tudor arch. Elsewhere there is a colonnade and several flat roofs to stone additions and gabled, half-timbered projections whose roofs feature decorative bargeboards and windows recessed under a bracketed beam.

On the main façade carved stone parapets flank the great entrance pavilion with its carved coat-of-arms label and framed cartouche. These extend on either side to small cross gables which conceal the eaves behind them and thereby add to the apparent height and importance of the whole façade. Opposite, across the driveway, are the tall, wrought-iron gates whose stone pillars are capped by giant sculptured urns. Extending in either direction from these gateposts is a high brick wall, inset at frequent intervals with heavy stone pillars. This encircles

the three-acre estate. A shorter wall, at right angles to the gateway, separates the main carriage drive from the great courtyard in which the garage building is located—a stone-trimmed, one-story brick structure, somewhat like an early stable with a flat roof which accommodates a flower garden and wide, arched doorways for the cars. Behind this is another single-story stone building whose parapeted top is sheltered in one corner by a separate roof on decorative wrought-iron pillars. At ground level, an arcaded passageway leads to the mansion's rear terrace and the garden steps, the latter illuminated by a pair of tall, iron lanterns.

In spite of the mansion's pristine look, the illusion of antiquity is scarcely lessened by a walk through the picturesque parklands, which are replete with sloping lawns and tumbling rock gardens buried under cascading multicolored flowers. In a pool beyond this rocky outcrop, at the end of a narrow peninsula, is a circular, stone and wrought-iron summerhouse, a thing of beauty with its seven clustered pillars, filigreed gallery, and a lacy ironwork roof curving upward to a gilded wind vane at its apex. A little farther on is a long and narrow lily pond terminating at the steps below an exquisite arched wellhead of white marble and exotic, patterned tiles that looks as though it came from some raja's palace in India.

A roadside view of
the Tudor manor house and
its half-timbered annex.

View from the island summer house.

The water-lily pond looking toward the white marble wellhead.

At the other end of this pool is a huge fountain supported by five life-sized stone elephants against a background screen of tall pillars connected at their tops by looping arches of decorative ironwork and punctuated at their bases by wide stone benches shaped from solid blocks of granite.

Such, then, is the exterior aspect of this Denver dream palace, seemingly untouched by the finger of time—until one steps inside. In preparation for this sally into the past, it should be mentioned that Mr. Minissale is Sicilian-born, steeped in the Italian love of art and music. A friend of the great operatic singers, from Caruso to Martinelli, and lover-*conoscitore* of things Italian, he has surrounded himself and his family with objects useful and otherwise that please his artistic soul: the warm and colorful decorative trivia and furnishings that remind him of the sunny land that gave him birth, without reference to the opinions of experts. The result has been the creation of interiors of individuality where the least costly item may be the most highly prized.

In the entrance hall all is color and life and drama. And everything is carved—the doors and doorways and ceilings—all of the decorating having been done under the eye and direction of Professor Antonio Romanelli of Florence.

Apart from this detailing, the feature of the foyer, once inside the bronze and iron outer door and the carved, bronze-studded inner one, is the magnificent Chinese rug patterned with bright-yellow and pink chrysanthemums on a black ground which covers much of the travertine floor. Here the walls are of Indiana limestone, the door surrounds of carved marble, and, if one can for a moment ignore the splendid Venetian tapestry whose colors harmonize with those of the rug, the extraordinary ceiling of embossed leather, the color of old parchment. Centered in this is a tall, five-armed bronze lantern.

Facing the entrance doors is an equally striking portal, with overdoor lights and carved marble screening, that opens into the drawing room. Beside this opening are massive pedestals on which stand a pair of giant-sized Sèvres urns adorned with classic scenes in pink, white, gold, and black, signed by Labarre. Above them, two black and bronze *torchères* cast a mellow light.

From this foyer the door to the right opens into the stair hall where travertine steps sweep in a huge circle up to the second floor, carrying with them the mahogany handrail on its iron balusters adorned by *repoussé*-decorated bronze plaques and floral studs. In the floor's center stands a marble fountain, a four-faced maenad supporting the upper basin, while the goddess of music plays the pipes above it.

Directly over the fountain hangs a Louis XV bronze lantern on a level with the stained-glass inserts of the tall stair windows. In this area, on the walls below, hang several well-known Italian paintings in gilded *repoussé* frames, and on the floor exotic plants in colorful vases of faïence, terra-cotta, and Majolica ware bring a touch of nature indoors.

In the rear wall of the stair hall are two doorways, one of them occupied by a pier glass which closes off the outer entrance to the butler's pantry, the other opening into the dining room. Between them stands a lovely Louis XV painted console table of a gay yellow, with carved and pierced decoration on its apron and legs, rope molding on the stretchers, and a marble top. On it stands a modern vase-based lamp, reflected in a cut-glass mirror hanging behind it. This antique mirror is of special interest, being made up of French beveled plate squares, with a border of diamond-shaped pieces. On either side of the lamp are a pair of French figurines in porcelain, with a bisque miniature of Salomé between them. The octagonal clock above the mirror is also an antique, the beveled case edged with mirrored strips and painted panels and the dial, brightly gilded, set in a field of marbled green. This is flanked by two three-light sconces that illumine the dial and much of the wall.

Turning back to the drawing room, despite the colorful furnishings, attention is immediately drawn to the fantastic coffered ceiling which is a mass of decoration in high relief and delicately gilded all over. There are three ceiling-high windows with decorative arches, inside which hang figured silk draperies in dark Burgundy and gold over handsome hand-embroidered fillet curtains.

In brilliant contrast are the deep-blue grounds of the two Chinese rugs dotted with pink, cream, yellow, and green flowers and the hearth rug of blue, red, pink, and beige, scattered over the old parquet floor. At one end of the room is a carved fireplace mantel in *verde antico*, its tall, gilt-framed mirror reflecting part of the handsome ceiling.

To one side of the mantel, a two-leaf door opening into the library has a carved panel set into the doorhead—a feature of several other doors in this area. On the other side of the mantel, a large painting depicts a garden scene dominated by a peacock's plumage, with a seated feminine figure floating in the background. Elsewhere on the green-tinted walls, *trumeau*-type mirrors depict French country scenes of the seventeenth century.

In this rich setting, the colorful Italian furniture of pale, gilded wood, carved and pierced, and supremely curvilinear, is decked out in deep red, deep blue, or pink silk damask; one of them is a *canapé à corbeille* type of sofa. The sole non-Italian piece is a Chippendale-style love seat in one corner, which is covered in a seventeenth-century tapestry. Behind this stands a fine French marquetry vitrine with ormolu mounts. At the other end of the window wall is a modern Steinway grand piano and, close by, a small, ebonized, bronze-mounted commode, while across the room a lovely French Empire bronze *guéridon* has griffins for legs. This serves to hold a globular lamp with circular translucent panes of yellow glass. A similar lamp has a yellow globe dotted with pink, blue, and red spots. The other small lamps in this room are varied in style, emphasizing

The Italian summer house on its peninsula in the pool.

Five life-sized elephants support a huge fountain set off by tall pillars and ironwork arches.

The foyer with its bronze-studded door and magnificent Chinese rug.

the eclectic nature of the furnishings in this palatial interior where the choice of pieces and accessories—pictures in particular—has been governed solely by the owner's individual judgment as to what is good and suited to his taste.

Another unusual yet highly impressive room in the mansion is the dining room which probably has no duplicate anywhere—including the setting. This is quite a large room, with all details in proportion—except the chandelier which is somewhat smaller than usual. The floor is a fine parquet, and quite old; the walls are beautifully paneled, each section exhibiting a faint, light-green patina and separated by a wide decorative band. The two end doors, one of them opening from the foyer, the other from the butler's pantry, are paneled, as are the stiles, and the overdoors have semicircular recesses framed within crested moldings that contain floral paintings. The panel surmounting the door to the drawing room is slightly different, having a circular painting set into it.

The baseboards in this room are exceptionally deep and molded top and bottom, while the room's frieze and cornice are heavily two-dimensional, enlivened with large dentils, egg-and-dart molding, and acanthus bandings below. The chimney breast, which projects about ten inches, has fluted columnar moldings to round off the corners close to the ends of the mantel.

The mantelpiece is of Italian marble, white with black veining, the opening covered by a Venetian wrought-iron screen of intricate design. Above the mantel is a tall, round-topped mirror set into a carved panel and finished with a crest consisting of an oval cartouche draped with tasseled cords and carved husks in imitation of the overdoors.

Opposite the fireplace is the great window, a splendid architectural composition of two slender, fluted, quarter-round pillars, topped by a double band of acanthus leaves in imitation of a capital, supporting an architrave decorated with a band of floral wreaths, all gilded to match the walls. In the recess behind the silk curtains are three swinging sashes with diamond-shaped leaded lights. The one smaller window close by is treated as a wall panel, enframed in gilt wood with the usual decorative band.

At each side of the great window stands a massive bronze *torchère* in the twisted columnar style of the Baroque period. This is topped by a Corinthian capital which supports a wide decorative ring of bronze inside which the light bulbs are concealed. Out of this sprouts a maze of gilded bronze leaves and flowers—the whole thing a tremendous Rococo *tour de force*.

In the rear transverse wall is a triple archway formed of white marble partly overlaid by the woodwork, which forms a panel in each pair of spandrels and substitutes the decorative cartouche and swags for each keystone. The sixth room surface to be noted is the decorated ceiling with light moldings in the Adam manner and fan-shaped fillets in the corners. This

Looking from the foyer into
the drawing room.

99

A glimpse of the travertine-floored conservatory.

also has a golden sheen, and from its center hangs an eight-light crystal chandelier.

The principal piece of furniture is, of course, the dining table, an extraordinary reproduction of an Italian original made in the United States by Italian craftsmen. Its magnificent oval base and top are formed largely of walnut veneer, but the decorative and unusual undulating apron is carved from the solid wood.

Equally striking in appearance are the chairs which were made of olive wood at the same time. These seem to be a compromise between a seventeenth-century Dutch design and the Italian Barochetto style, though the usual painted flowers are replaced by the elaborate pierced carving of the splats and crestings, and the ball-and-claw foot by the scroll. The seats are covered in a pale-blue and silver silk brocade. Closely conforming to this style is a mahogany tea table, with an undulating top and decorated cabriole legs, which supports a great silver urn.

The gilded mirrors of a most unusual shape are French, including one over the sideboard which has a Rococo-style carved frame. The remaining pieces are vitrines for the storage—and display—of silverware. These actually are somewhat somber notes in a room so full of elegant life and sparkle.

Through the dining-room arches can be seen the travertine-floored conservatory with its many round-topped windows set

OPPOSITE TOP: The stair hall with its travertine steps and central marble fountain.

OPPOSITE BOTTOM: Through the dining room arches is seen the equally dazzling conservatory.

One corner of a guest bedroom.

in the stone walls. Here are the large, glass-topped iron table and the wrought-iron chairs, all in a pale green except for the yellow cushions. In addition there is a flowered chaise longue, an Italian-style iron chair, and cast-iron bench, while an arty atmosphere is supplied by a fanciful white-marble statue in one corner and illumination by two seven-candle floor stands of wrought iron.

Upstairs is to be seen some of the most beautiful woodwork in this fascinating house and some of the finest craftsmanship in metal. In the second-floor hall the parquet pattern changes to a simple English style, and the winding stair rail is seen at close quarters, together with its brass-decorated iron balusters. Far more exciting, however, is the filigree work that forms the gates leading to the family's private apartments; this must be among the most beautiful designs ever executed—the whole metal curtain a poem in black and gold.

The blond, mahogany wall panels, too, are worthy of mention, framed as they are in veneer banding. The central boards are manipulated to produce a pleasing grain pattern against which a modern tall clock in oak and polished brass appears at its superb best, and the bronze and marble console table, with a gilt-bronze *appliqué* above it, is perfectly at home.

In this mansion there are many other rooms, but enough have been shown to indicate opportunities for exotic treatments of which full advantage has been taken in creating a wholly Latin atmosphere in a modern American adaptation of a wholly English house, so that, in passing through the portal, one is at once transported from the world of Shakespeare into the world of Dante.

At the stairtop stands this elaborate bronze gate of the family's living quarters.

103

OPPOSITE: The drawing room with its fantastic coffered ceiling.

THE HAY HOUSE

Italian Renaissance of the 1850's

The Renaissance-style
Italian villa.

While honeymooning in Italy in 1851, William Butler Johnston and his bride became so enamored of the Renaissance-style architecture that they decided to build an Italian villa of their own in Macon. Accordingly, they arranged to import certain essential materials—other than bricks which were to be made on the site—together with a number of Italian artisans to work on the construction and decoration of the mansion under the direction of their architect, James B. Ayers. This project, because of delays engendered by the Civil War, was to occupy seven years, though the Johnstons were able to move into the house well before the end of hostilities.

With the building well under way, furniture for the major rooms was ordered from Italy. Unfortunately, by the time it was ready for shipment the Southern ports were under blockade by the Federal navy, and only one set of furniture—that for the dining room—escaped seizure. This still remains in the house despite the fact that the property was first inherited by the Johnstons' daughter, Mary Ellen, wife of Judge William H. Felton, and sold by their grandson in 1924 to another prominent citizen, Park Lee Hay. It was he and his wife who acquired these pieces, together with the rest of the splendid furnishings that help make this house the showplace that it is today. The building itself has the distinction of being unique in the United States—a decidedly impressive three-story brick structure above a raised basement, topped by a massive octagonal stone cupola with alternating arched and *œil-de-bœuf* windows, scroll brackets, and a balustraded octagonal lantern with a walkway, like something by Borromini himself.

The main façade of the central block of this mansion is definitely Romanesque, with tall, arched, and keystoned main-floor French windows and wide double doorway beneath a pillared portico, whose floor of patterned marble curves outward at the center to shelter the top step of the balustraded marble stairway which fans out to twice its width at the bottom. The flat roof of this single-story portico, whose columns are a

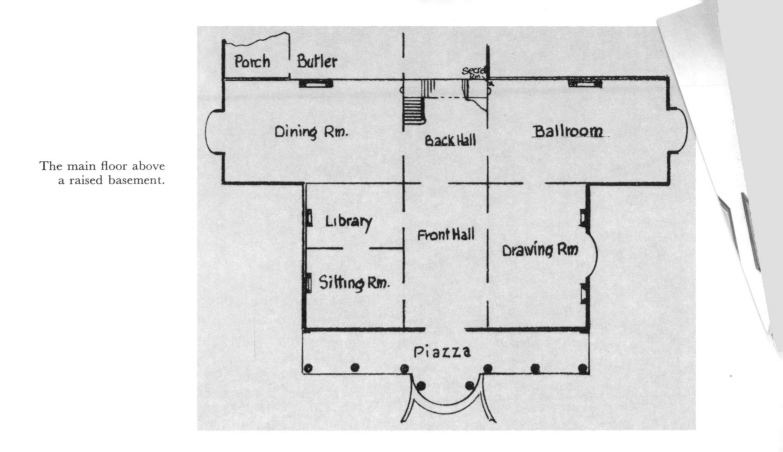

The main floor above
a raised basement.

Rococo adaptation of the Corinthian order, is also balustraded to form a balcony the full width of the building. This is accessible from any of the five second-floor French windows, four of which are sheltered under either a triangular or segmental pediment. The exception is the central window which is grouped with its sidelights to form a single architectural unit spanned by a composite pediment whose tympanum is ornamented by carving in relief. Over this group of French windows extends a band of stone molding which forms a stringcourse right around the central unit. Between this and the modillioned cornice are spaced five *œil-de-bœuf* windows, enframed by carved stone rings.

The roof is a low-pitched, hipped type covered in terneplate with standing joints, and at the bottom of each side slopes a pair of decorative brick chimneys, complete with tall earthenware pots, adding a further exotic note. On either side of this main unit is a single-story wing, one room deep, with a glazed monitor roof capable of flooding the rooms below with light. The front and rear façades of these wings have modillioned cornices and a stone band above the windows, all of which, in this case, possess triangular pediments which help give the three sections a feeling of unity.

Despite their Victorian air, the twin front entrance doors are probably unique both in size and design. These huge portals are said to weigh 500 pounds apiece, being several inches thick and heavily studded, with deep moldings to the round-cornered panels which feature carved lion heads in high relief at the center and upper panels of etched glass. The locks are huge, and the great silver-plated hinges are still in perfect working order after 110 years of service. Their design and construction, which symbolize the qualities of the entire structure, are sufficient to indicate what may be expected

Etched glass, lion heads, and silver-plated hinges
distinguish the front door.

105

A picture of Lord Byron caps this stained-glass window.

The hall's Carrara-marble floor and decorative plaster work.

inside: the wide front hall floored with white-marble squares and dotted with black cabochons; a heavy central archway resting on large and intricately modeled corbels that look more like Rococo dripstones; its elaborate plaster frieze and cornice and fancy paneled ceiling, all in keeping with its amazingly thick walls and contributing to an air of solidity that the sturdy interior detail does little to disguise.

Beyond the second archway—this one supported on pilasters—is the stair hall, where the carved mahogany staircase provides the same baronial atmosphere in spite of the white walls and the stained-glass windows above the landing. But here also is a smell of intrigue. At one side of this landing, above the dado, is a deep, wood-lined niche displaying a lovely two-handled porcelain vase. At the other side of the landing is a similar niche and a second vase, but this time the round-backed, arch-topped casing can be swung aside to reveal a secret room that was made use of by the Confederacy toward the end of the Civil War.

In the front hall, most of the floor is covered by two colorful Oriental rugs which form a connecting link between a pair of tall-back "lolling" chairs and twin Queen Anne armchairs; a carved mahogany library table and a tall cabinet-on-stand, painted with mythological scenes from Dante's *Inferno*, form a pleasant grouping with tall, mirrored wall sconces and a metal-framed looking glass of unusual design. Other luxurious touches are supplied by a scroll-topped mahogany tall clock and a marble pedestal on which stands a fine Meissen vase.

Even more elaborate and artful than these halls are two of the more formal rooms beyond, each of them lavishly decorated with sculptured white marble, carved and molded wood, and intricate stuccowork enlivened by loops and whorls, quirks and latticing, all in the best of taste. These permanent features, though lively in themselves, were obviously intended as backgrounds to enhance the furnishings instead of overpowering them, supposedly in a period when the Baroque was giving way to the milder Rococo. For the most part, this objective was achieved as can be observed in both the ballroom and the drawing room.

The ballroom, which occupies the right-hand, single-story wing, was originally intended as an art gallery for the display of the Johnston collection of paintings, porcelains, and sculpture, though actually the wall spaces are quite limited because of windows on both sides of the room, the fireplace, and the wide archway of the entrance. On the other hand, the ceiling is 30 feet high and deeply coved above an extraordinarily detailed modillioned cornice. This 50-foot-long coved area centers on a tremendous skylight with stuccoed borders and decorative panels from which depend a pair of huge Waterford-crystal chandeliers that sparkle even in the daylight, thanks to the natural overhead illumination.

Centered in the outer end wall of the room is a deeply recessed window, set within an arched opening, and a short length of barreled ceiling with sunken panels in the plaster. This recess is flanked by attached, fluted columns with the same pseudo-Corinthian capitals as the portico that gives them individuality. Within this area is a lovely sculptured figure of the Biblical Ruth in Carrara marble by Randolph Rogers of Rome. This is mounted on a 4-foot marble pedestal against a pyramidal backdrop of golden draperies tied back to reveal ethereal mesh curtains, which seem to glow in the filtered sunlight.

Within this recess and around the rest of the room is a dado of sunken panels, stopped against the shallow chimney breast which houses an arched Victorian fireplace in white marble. Facing this across the room is an acanthus-molded entrance arch on fluted pilasters, with tall sliding doors of hand-carved rosewood and upper panels of etched glass. All of the rest is applied decoration which sustains the mood—a huge gilt overmantel mirror with a cresting almost Baroque in its exuberance; Roman-arched window tops hidden beneath shallow, concave gilt pelmets from which hang gold-fringed central swags in blue silk, over pairs of pinkish-tan ones that form cascades down the sides, revealing floor-length, tieback draperies in gold.

More than half the length of this room is covered by a huge Savonnerie rug with a giant pattern in gold, black, red, and royal purple from the royal factory in Vienna. This looks every bit as rich as the remarkable copy of a Louis XV marble-topped *bureau plat* in boullework with ormolu sabots and chutes

A "secret room"
on the stair landing.

The ballroom mantel and
original Parisian vases.

standing alongside it. Over this desk is a painting of a cathedral by C. J. M. Whichelo (1805–1865), which was considered the most valuable picture in the Hay collection. Across the room, over a console of white mahogany, which has camel legs and a *verde antico* marble top, hangs another large painting: Jules Lefebvre's *Lady of Sorrento,* once a feature of the Cornelius Vanderbilt collection.

In fascinating contrast with the ballroom is the drawing room, which extends the full length of one side of the mansion. This contains the most elaborate ceiling in the mansion—and probably in the country—the work of the stuccadors brought from Italy more than 120 years ago. The detail of the frieze and cornice is almost beyond belief, comprising a series of female heads in alto-relievo, each enframed by an intricately modeled cartouche, bordered by swirls, loops, and drops every eighteen inches or so, each separated from the next by a swag or fruit and leaves, and the whole thing suspended from a continuous rod by a leafy creeper. In addition, every cornice has its tiny molding of oak leaves; set deep within this border is the flat ceiling area adorned with huge rectangular and oval panels, each with a complicated rosette at its center, the two outer ones each suspending a crystal chandelier.

This room also has a wide, arched window recess with paneled pilasters and reveals, each of the bay's three round-topped windows forming an arch within which the silk tieback draperies are hung. Three similar archways, two opening into the hall, the other leading to the ballroom, have twin sliding doors of carved rosewood. The windows too are round-topped, though the arches are draped with a double swag, inside which the sheer curtains are hung. All the woodwork here is painted a creamy white, the walls being covered in an apricot-colored flock paper, the design nicely proportioned to the room's stately dimensions.

Over the twin, white-marble mantels are a pair of rectangular gold-framed mirrors with colonnettes and a swag-ornamented frieze under wide Rococo cresting and terminal scrolls whose elegance is timeless. The *garniture de cheminée* consists of Sèvres, Meissen, and Capo di Monte vases and urns, and to one side, on a marble columnar stand, is a blue Meissen urn.

Much of the furniture duplicates that of the ballroom, though the upholstery is in colors calculated to blend with that of the carpeting. The oval-backed canapé, for example, is in blue-striped pink silk, while the oval-backed chairs are covered in pink satin. Equally at home here are an oval table by A. C. Boulle (1642–1732) and the round coronation table of Louis XVI in ormolu and ebony inset with paintings on porcelain of the court ladies. The two most interesting paintings are Angelica Kauffman's *Hebe,* posed for by the Countess of Carlisle, and a portrait of the Comtesse de Chevigny by Jean Baptiste Van Loo (1684–1745).

The only wholly Italian interior is that of the dining room. This has a high vaulted ceiling totally covered with recessed

In the drawing room is the Hay House's most elaborate ceiling.

A Tiffany stained-glass door leads to the butler's pantry and kitchen.

OPPOSITE: This dining room possesses the mansion's only wholly Italian interior.

Portraits of Mr. and
Mrs. P. L. Hay hang in
the French sitting room.

At one end of the ballroom
is this revolving statue
of the Biblical Ruth.

oak paneling, each inset panel framed by elaborate carving
and centered on a drop ornament, the single exception being
the central panel which forms a skylight from which hangs a
crystal chandelier. This dark ceiling is supported by heavy
curved beams with paneled soffits, resting at each end on a
huge wooden, scroll-shaped corbel. The whole of this woodwork is painted green, enlivened with touches of blue and gold.

Though the walls are finished in a light beige, their reflec-
tiveness is considerably reduced by the massive carved-oak
hoods and molded frames of the round-topped windows and
doors, by the archways leading to the ballroom and the garden,
and by the almost ceiling-high wooden overmantel. The mantel
itself is flanked by built-in sideboards, and the garden window's
pilastered archway by vitrines over which twin blind windows
are set with Tiffany glass and mirrored squares. More Tiffany
leaded glass fills the skylight and the door to the butler's
pantry. Above the heavily paneled dado, the plaster walls are
tinted beige.

This floor is of parquet, largely covered by an Italian rug
which further tends to darken this somewhat overpowering yet
interesting interior, and the tieback window draperies of red
brocade, set within the paneled reveals, still further reduce the
available light. All of the furniture was made in Italy, including
fourteen chairs with upholstered seats and backs; a massive
extensible dining table with bulbous legs, carved apron, and
heavy central pedestal; and a comparatively simple sideboard
whose turned legs match those of the chairs. This is adorned
with a silver urn and a pair of crystal lusters.

Though there are many other interesting rooms in the mansion, none could display to better effect than these three the
possibilities of adapting a fine Italian Renaissance-style villa to
nineteenth-century American living. Today the mansion is
preserved for posterity by the heirs of Park Lee Hay through
the Hay Foundation as an exemplar of enlightened Southern
taste in antebellum days.

OPPOSITE: Originally an art gallery, the ballroom is lighted
by a huge pair of Waterford crystal chandeliers.

Newport, Rhode Island

MARBLE HOUSE

American Renaissance in Excelsis

In a city of splendid mansions, Newport's Marble House stands out as a breathtakingly lovely example of Renaissance architecture. Built in 1889 to 1893 by Richard Morris Hunt for William K. Vanderbilt, the structure, so reminiscent of the Petit Trianon, is palatial in feel if not in detail both inside and out, though on occasion it seems that certain of the more lavishly decorated and furnished interiors tend to be not merely grand but grandiloquent.

Much of the furniture, whether of oak, walnut, or metal—always in period style, though wherever possible accenting the Rococo—was made expressly for the rooms in which it was to be used, the bronzes especially designed or copied from world-famous originals, and everywhere there is gold and polished marble, elaborately painted ceilings, and tremendous shimmering chandeliers, creating a fabulous display that would have delighted the Sun King himself.

From the entrance driveway, the principal façade of this U-shaped building, which faces west, is an exquisite confection of light-gray limestone with a central portico of tall Corinthian columns, the walls on either side elaborated with fluted pilasters, all beneath a full entablature whose deep modillioned cornice is surmounted by a continuous stone balustrade extending right around the roof. The fully developed Corinthian capitals of the columns and pilasters stand out boldly against a background of smooth walls empaneling the round-topped, main-floor double French doors and severely outlined rectangular upper windows, as well as the huge bronze gates below them in the shadow of the portico.

Up to this main entrance sweeps a semicircular, inclined driveway, bordered by heavy stone balustrades terminating at the portico steps. Here the outer balustrade gives way to a richly decorated bronze railing, from the front of whose projecting stone base three carved masks spout water into a half-round, stone-rimmed pool. Along both sides of this carriageway loop are massive metal lampstands mounted on the balustrade

The entrance driveway with its heavy balustrade and bronze railing.

OPPOSITE: Smaller than most rooms but just as rich is the library with its great chandelier and allegorical ceiling panels.

113

The principal façade of gray limestone with Corinthian columns.

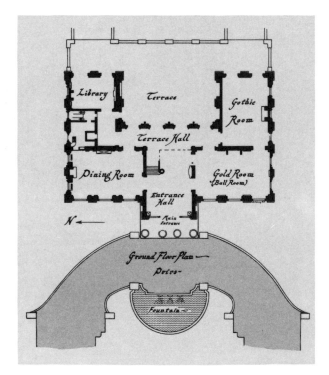

The ground-floor plan.

plinths. These form decorative features by day and, after dark, illuminate the whole area within its ring of tall trees.

On the east façade, which faces the ocean, the open space between the mansion's two wings forms a wide terrace that is equally lovely in a different manner. The three eastern faces of the building are punctuated with the same two-story Corinthian pilasters, and the spandrels of the window arches are carved with the four signs of the zodiac, while the panels between the pilaster capitals display garlands of fruit and flowers in high relief. Below, on the marble terrace itself, there are two marble benches supported by satyrs with musical trophies on the panels between them, and against the central court wall stand three marble terms from France representing, in turn, a satyr, a goddess, and a nymph. Different again is the south flank of the mansion, where each of the five bay arches is pierced by a lunette into which is set a relief carving of some mythological figure.

Fascinating as these all are, the front entrance outdoes them with its 25-foot-long bronze grille, decorated in gold, including a mask of Apollo centered in a sunburst (a favorite device of Louis XIV) and cartouches engraved with the William Vanderbilt monogram. Inside the pair of 6-foot-wide doors, the entrance hall is equally superb, the 20-foot-high walls and the floor being faced with yellow Siena marble, the ceiling decorated with elaborate arabesques in stucco, and the walls adorned with huge Gobelin tapestries, woven in the shops of Pierre François Cozette, set in gilded frames. On either side, marble-lined doorways open into the ballroom and the dining room, and, straight ahead, is the huge stair hall where the ceiling suddenly rises to 40 feet from the floor, and a huge stunning bronze lantern hangs from it in the rectangular stairwell. The railing of the marble stair is of bronze ornamented with gilded plaques and the Louis XIV sunburst, duplicating those on the entrance grille.

There is little furniture in these two sections of the lower hall, but what there is is exquisite. The pieces include four Louis XIV-style open-arm chairs of gilded bronze upholstered in red velvet; a Louis XV-style *bureau plat* heavily ornamented with bronze *doré;* and a somewhat smaller nineteenth-century boulle cabinet. In the corners are tall bronze lampstands ornamented with gilded dragons inside heavy volutes, and a bronze fountain between the Gold Room doors—all by J. Allard, *fils*—over which is a heavy mirror, enframed in etched glass in the Venetian manner. Beyond these, at the rear of the hall, stands a tall boullework cabinet supporting an elaborate French mantel clock in bronze.

Up the first flight of stairs, at mezzanine level, are two round-topped French windows with heavy draperies and intricate pelmets of cut velvet. On a wall bracket between them is a marble bust of Louis XIV as a young man—a copy of the seventeenth-century original by Giovanni Lorenzo Bernini (1598–1680)—while over the arches gilded figures hold white-marble portrait medallions of the house's designer, Richard Morris Hunt, and the chief architect to Louis XIV, Jules Hardouin-Mansart (1645–1708), by Karl Bitter, the noted New York sculptor.

On the second-floor landing the wainscot and door trim are of Siena marble, but the upper walls are marbleized in order to duplicate the appearance of the lower walls and staircase. The most striking feature here is the coved ceiling whose corners are enlivened with heroic-sized figures in red-and green-tinted gold, while at the ceiling center gods, goddesses, and cherubs disport among the clouds in the manner of a Tiepolo painting. Much more realistic is the wall-hung portrait of Louis XV as a youth, a painting attributed to Jean-Baptiste Van Loo (1684–1745).

The most lavishly decorated of all the Marble House interiors is undoubtedly the ballroom—inevitably referred to as the Gold Room when it was used for other purposes than dancing, at which times the floor of diagonal herringbone block parquet would be covered with a beautiful Savonnerie carpet. This sumptuous room is a riot of overwhelming detail in shimmering gold, endlessly repeated in oversize mirrors—surpassing even the royal bedroom or the Hall of Mirrors in that world-famed Neo-Baroque fantasy, Ludwig of Bavaria's Linderhof of 1874. The high, gilded walls, with their elaborately detailed frieze, accommodate a series of wide arched doors and windows, huge red, green, and gold panels carved in relief; wall-high mirrors; and a great haunched *fleur de pêche* marble fireplace on which are seated heroic-sized bronze figures, representing old age and youth, upholding tall candelabra and tied together by bronze swags centered on a mask of Bacchus. Above the mask,

Exquisite Gobelin tapestries are featured in the entrance hall.

In the stairwell hangs a great bronze lantern.

The most elaborate interior is the ballroom, glittering with gold under a huge allegorical painting.

a glass terrestrial globe marks the time by revolving around a sphere representing the universe.

More imposing than anything else in the room are the two large chandeliers, their myriad branches swathed in gilded leaves, while a ring of life-sized *putti* with trumpets encircles the base. In the midst of all this glitter, the one serene spot is the central ceiling panel: a huge allegorical painting reminiscent of seventeenth-century work, its gilded frame surrounded by white stucco panels and gilded wreaths in the corners, over a deeply coved, gold-ornamented double cornice. The doors and windows are draped in brown and gold cut velvet, duplicating the upholstery of the gilt-bronze framed wing chairs and sofas, and trimmed with silk fringe studded with rhinestones.

Quite subdued, in contrast with the ballroom, is the Gothic Room into which it opens, even though the entire ceiling, braced by curved corner struts resting on corbels, forms an elaborate tracery of beams and panels. The door and window embrasures, and most of the wall areas, are sheathed in intricately carved wood paneling between ceiling-high pinnacles. All of this endows the room with a highly ecclesiastical air, an impression fortified by the three-dozen angelic musicians spaced around the frieze.

Equally elaborate is the Gothic stone fireplace—a replica of that in the Jacques Cœur house in Bruges—with its arched, highly detailed, and canopied niches and its crocketed finials flanking three panels, in which rapt onlookers are depicted as gazing out upon the scene while helping themselves from a bowl of fruit or playing chess. A knight in armor and a habited monk support the mantelpiece. Even the crenellated chimney breast with its blind machicolations—common interior decorations in the early centuries—suggests the abbey rather than the castle.

This room was planned to house a collection of Gothic artifacts long since dispersed and was decorated accordingly, the great coved ceiling painted in polychrome, the upper walls covered in red damask, and the windows of stained glass arranged to provide extra illumination to the interior as required. From the ceiling there still hang twin chandeliers of dull bronze, while all around the room four-candle wrought-iron floor stands and similar candelabra on the table are available for use as required. Today the ceiling, the fireplace, and all the upper walls are white plaster, except for spaces covered by velvet armorial banners and one seventeenth-century Flemish tapestry.

One of the Gold Room's lavish panels.

Mrs. Vanderbilt's study.

The floor is of heavy pegged planks, adding still more wood tones to set off the colorful tapestry coverings of the tall-back Louis XIV armchairs and the velvet of the fireside stools in a room far more restful—and authentic looking—than it must have been in its gaily colorful nineteenth-century dress.

Almost as lavishly decorated as the ballroom is the dining room, but here the mood is far more restrained and comfortably relaxed in spite of its obvious richness. Although the profusion of gilt ornamentation is much in evidence, its livelier areas are concentrated at ceiling height. At lower levels the vast expanses of pink Numidian marble are far less aggressive—and oppressive—their softly rich tones bespeaking true quality. On the ceiling, too, a huge painting depicting some mythological happening among the Olympians seems to have a quieting effect upon the gilded stucco trophies depicting hunting and fishing paraphernalia, garlanded with sea shells, and trophies in high relief spilling out of the corners.

In the wall angles beside the fireplace, the doorways, and the windows are marble pilasters of the Corinthian order, with bases and capitals of gilt bronze, those bordering the chimney breast overlapping to form an overmantel panel adorned by a gilt-framed portrait, attributed to Pierre Mignard (1610–1695), of Louis XIV as a young man. Below is the bronze-ornamented marble mantel with a mask of Hercules at the center, a design inspired by that of the Salon d'Hercule at Versailles. Between

Consuelo's bedroom.

William K. Vanderbilt's bedroom.

OPPOSITE: The dining room with its portrait of Louis XIV as a young man.

each pair of pilasters around the walls is a multibranch bronze *appliqué* in the form of a cluster of pennants, spears and other arms, and a plumed helmet in the center. Near the hall doors are tall, gilt-bronze lampstands adorned with Apollo masks and goat heads, topped by flower-filled vases. On the polished mahogany top of the great dining table rests a pair of gilt-bronze candelabra, each of which holds ten tapers.

All of the chairs, stools, and side tables in this room are of Louis XIV design, the impressive seated pieces made of cast bronze and upholstered in rose-pink cut velvet. Inside the round-topped window recesses are braided and fringed pelmets and draperies of crimson cut velvet, with cord tiebacks tasseled in gold. These look none the worse for being paired with a floor of the same herringbone block parquet as the Gold Room, a style undoubtedly chosen because the pattern does not emphasize perspective.

On the second floor is Mrs. Vanderbilt's bedroom which,

The "austere" Gothic Room
with its bare floor.

Mrs. Vanderbilt's bedroom swathed in peach damask.

though exquisite in every detail, happily declares its owner not to have been so enamored of Louis XIV as to wish to emulate his retiring and arising ceremonials. Hence there is no balustrade around the bed. The bedroom, nevertheless, leaves no area of wall, floor, or ceiling unadorned with abundant detail in the luxurious Rococo style. The walls are covered in a peach damask of an arabesque pattern, and the curtains are of the same material, with pelmets of carved and gilded wood depicting a bonneted female mask at the center, with a cherub on each side grasping a triple floral festoon, under which are draped fringed swags of the curtain material. These are quite in keeping with the slightly coved ceiling which has elaborate borders and panels of stucco decoration around a central ceiling painting of the goddess Athena and four smaller corner panels each enlivened with a single cherub. All the window recesses and the dado are decoratively carved, as are the panels set at forty-five degrees across each corner from dado to cornice. Similar vertical panels form pilasters each side of the *fleur de pêche* marble mantel, which not only has an elaborate shell carving at its center but balances an almost life-size bronze figure on each shoulder, holding high a pair of gilded candelabra.

Not quite so embellished are the rest of the rooms, though most of them have stucco-decorated ceilings and rinceau friezes. Daughter Consuelo's room, however, has a paneled ceiling and wainscoting, a deeply canopied four-poster bed, and a Tudor fireplace, all decidedly unfeminine and somehow prophetic of her future as wife to the Duke of Marlborough and chatelaine of Blenheim Palace, one of England's noblest houses.

Mr. Vanderbilt slept here.

121

Middletown, Connecticut

THE ALSOP HOUSE

A Neo-Classic Mansion Unique in Design and Decoration

The Alsop House in winter.

As an architectural composition, the Alsop House, of which Ithiel Town of New Haven is believed to have been the architect, is unique. Part of this distinction lies in the fact of its being decorated, both inside and out, with *trompe l'œil* representations of statuary-occupied niches, while its principal interiors—both walls and ceilings—are embellished with lively paintings which, by the skillful use of oil colors on dry plaster, simulate frescoes, the whole series being the creation of some unknown artist, probably German, in the eighteen forties. Regrettably, some of these on the exterior have had to be reproduced, since the originals—still preserved but not exhibited—were well beyond restoration.

For these reasons the house itself, which is of brick stuccoed over and painted a pinkish tan, is acclaimed as being one of the most notable architectural achievements of the pre-Civil War period, classic in concept, though entirely devoid of columns or pediments, outside or in. Built in 1838 by Richard Alsop IV for his widowed mother, it remained in the family until 1949—although not occupied for many years—when it became the property of Wesleyan University.

That this is something different from the ordinary neoclassical style of mansion of that period is immediately apparent. All of the exterior walls are quite plain and smooth, including the three giant rectangular pillars that support the somewhat lower roofs of the setback foyer structures, which form short wings attached to each side of the main block.

The main façade, two full stories above its flagged terrace, exposes five large, almost ceiling-high windows (two down, three up) beneath a decorative frieze under the widely projecting bracketed eaves of an almost-flat roof. At stringcourse level, between lower and upper windows, a low-pitched shed roof covers the three-sided terrace and is upheld at equal intervals by black-iron, treillage-style supports instead of solid posts, the roof itself being crowned by a continuous, decorative iron railing.

The wall decorations here consist of a window-high *trompe l'œil* painting between the two lower windows, depicting a statue of Erato, the muse of love poetry, set in a rounded niche, companion to those of Juno and Victory in the foyer and morning room wings, and fully in keeping with the swag, drop, and patera decoration of the frieze, in highlighted gray, which is carried right around the building. The only decorative addition to the windows is the use of wood trim in the reveals, with wood panels across the top carved in low relief and supported by board-thick decorative consoles at the sides, the whole ensemble painted a pale yellow which accents the voids.

Of all the house interiors the most striking is the stair hall, whose walls are painted to represent smooth ashlar blocks invested with a golden tinge—a mottled hue that becomes more evident in the walls of the lower hall. One of these walls is rounded to receive the lower, curving end of the staircase. At the top of this hollowed wall, the second-story ceiling opens into a circular dome, from the center of which hangs a bronze-banded, three-candle, bell-shaped lantern of clear glass that serves not only to illuminate the staircase but also the upper wall with its *trompe l'œil*, statuary-filled niches and, above them, the enormously detailed frieze, as well as the beribboned laurel wreath that garnishes the interior of the dome's base. One of these *trompe l'œil* figures is a copy of a Canova dancing girl, but the other has not been identified and may be original with the painter.

Almost equally a work of art is the double sweep of the stair rails which suddenly become horizontal to take in a length of landing area halfway to the ceiling. Another unusual feature is the installation of a second handrail against the wall, while

The plain exterior walls are enlivened by *trompe l'œil* statuary.

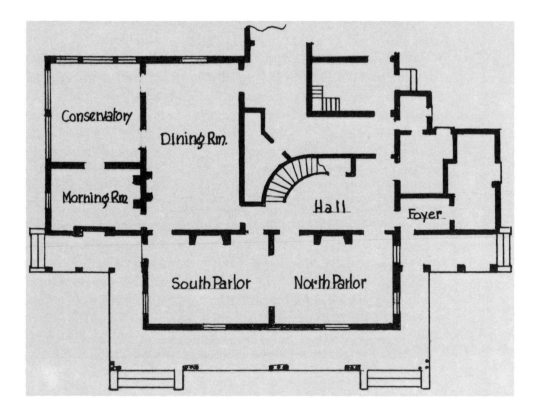

Floor plan of the original house.

The sweeping principal staircase emphasizes the main floor's ceiling height.

The staircase with its *trompe l'œil* statuary and hanging dome lantern.

the panels enclosing the space beneath the stair sweep around a lovely Carrara marble bust on a black-marble pedestal. Other decorative features of the stair hall include painted overdoor panels of drop-ended swags and paterae, duplicating those of the mansion's exterior frieze. The hall's furniture is naturally limited to an antique mahogany tall clock; a mirror-backed Regency console table with a marble top and lion monopodia in the Egyptian taste, over which hangs a Federal-style mirror of uncertain origin, and on either side, a pair of carved, saber-legged chairs of the same period.

One room in which art works have been called upon to complement the fresco decorations is the dining room, where the vast areas of plain wall space in a pinkish beige beneath a formalized brown-on-yellow frieze form a perfect background for a whole series of gold-framed Piranesi prints. In this room the tall casement windows and floor-to-ceiling glass doors to the conservatory are draped with frieze-high curtains of a heavy cotton fabric in pale green with pink and gray tassels—a perfect complement to the white ceiling and gray woodwork which includes four normal-sized doors. The fireplace mantel of white marble is quite plain; the facing is a mottled gray, and over it is one of the larger Piranesi engravings, while the garniture consists of an Empire clock and a pair of candelabra in porcelain and bronze, all quite in keeping with the French Empire-style crystal chandelier.

On this floor is a large Oriental rug in blues, reds, and beige, which forms a noncommittal background for the beautifully carved, birdcage-pedestal type of dining table of Phyfe design and shield-back English Sheraton chairs in a striped fabric of green, pink, and gray. The high-back mahogany sofa—another Phyfe piece—wears a green-and-gold cotton brocade in a medallion pattern. Some beautiful wood grain is also introduced by a silver-laden Sheraton sideboard of inlaid mahogany; a round-topped, eighteenth-century mahogany Hepplewhite card table, and a Sheraton satinwood-inlaid, serpentine-edged, mahogany card table, the two latter used to display antique porcelains in green and white.

The two most fascinating rooms, however, are the north and south parlors, which are separated only by wide, sliding doors so that their interiors must, of necessity, be quite compatible when they are thrown together. For this reason the wall decorations are duplicated, as are the ceiling designs, except for minor differences of detail. Duplication actually is confined to the Greek-key frieze and the paneled borders of the ceilings—end-pointed strips which are largely gray and light blue, with dark-blue arrowheads in the corners, together with a frame of floral swags and a central, eight-pointed panel formed of superimposed squares in blue and gray into which a circular painting is set, ringed by a half-round molding in gold.

In the north parlor this central ceiling painting depicts Apollo in his cloud-borne chariot pulled by prancing horses; in the south parlor Zeus is seen in his two-wheeled car drawn

The morning room with its painting decorations.

In the south parlor decoration differs only in detail from that of the adjoining north parlor.

by eagles, both frescoes copied from originals in the Vatican. Outside the swag borders the differences consist merely of wreath-bearing angels in the corners of one and trumpet-flourishing wingless sprites in the other. In both rooms the overdoor panels with formal floral designs are duplicated in a larger size between the window tops, while over each mantelpiece a frieze-high panel carries at its center a painting of a flaming bronze torch flanked by cupids, all mounted on an acanthus scroll. Similar panels on the walls are content with a single ethereal goddess or stylized urn with flowers and birds, all copied from prints of Raphael's *Hours*.

Although the ceilings and wall panels contribute their quota of color—soft gold, pink, light and dark blue, and orange—considerable warmth is added by the window draperies of Burgundy silk, their deep, pleated swags, gold-tasseled, hung over wreath-centered gilded brackets to form short cascades from which descend tieback, floor-length pongee glass curtains. Each room is also enriched with modern carpeting of pale gold that sets off the polished mahogany of an eclectic assortment of furniture pieces—the fine French Empire secretary-bookcase exposing fine Capo di Monte pieces behind its glass doors; the straight-legged version of a Late American Chippendale wing chair in a Burgundy cotton weave—a color much used in the wall paintings; a stately Regency *bergère* in a red-and-white-striped material on gold; lively Empire armchairs in a red and gold diamond-patterned brocade; side chairs of the same period with beautifully carved splats; a Phyfe roll-arm sofa in dark-blue velvet; and a beautifully inlaid English Regency sofa-table. The prize piece is a slant-top secretary-bookcase occupying the wall space between south parlor windows (an original Alsop possession made by the famous Townsend brothers of Newport, Rhode Island, during the Revolutionary War), unless one is a musician to whom the little eighteenth-century square pianoforte is of more abiding interest. The artificial lighting here is still provided by whale-oil lamps or Early Victorian candelabra converted to oil—with the exception of a tall green-bronze Tiffany lamp in a Pompeiian design which adds an exotic touch.

Quite different from the principal interiors is the little morning room, where the wall decorations and ceiling painting are light and airy suggestions of openings in the surfaces, spanned by fanciful iron tracery along which blooming creepers climb and on which perch a variety of well-known local birds—a theme very much in keeping with the morning sunlight which so often floods the room, even though the tall windows are shielded by eminently translucent silk draperies of a pale gold.

The south parlor ceiling shows
Zeus in his eagle-drawn car.

The dining room's silver-laden
Sheraton sideboard.

The stair hall's antique tall clock, Regency console, and Federal mirror.

An original Alsop secretary-bookcase in the south parlor.

The furniture here, too, is of mahogany and exceptionally sturdy. There is even a small fireplace of white marble which adds to the brightness and a subdued rug in a pattern of floral squares that calls no attention to itself. Larger in scale and more pungent in pattern is the fabric covering of the Sheraton-style Phyfe sofa whose huge laurel wreaths are white on green. Next to it is the far less flamboyant Sheraton lyre-back armchair with its heavily carved cresting, which looks perfectly at home with the saber-legged Empire side chairs that are duplicates of those in the south parlor. Another fine piece is the French Empire library table with end slides, a green leather top, and gilt-bronze trim, cheek by jowl with a lovely French cabinet which has a tall brass gallery and a gently rounded front, whose doors are inlaid with darker wood which, when closed, looks for all the world like a blind Gothic arch.

As in the rest of the rooms, the furnishings here are of the highest quality, though eminently practical and well suited to the beautifully decorated interiors which, together with the architectural excellence of the exterior and its extraordinary embellishments, endow the entire residence with a pleasing individuality it would be difficult to match.

OPPOSITE: Gold-framed Piranesi prints complement the dining room's fresco decorations.

New Castle, Delaware

THE READ HOUSE

Federal Grace Plus Georgian Charm

Few mansions of any period are as attractive as the Read House, which seems possessed of both inherent dignity and friendly charm—conflicting attributes arising from its graceful proportions combined with sturdy construction and elegant decorative detail both outside and in.

The building of this solid post-Revolutionary brick house was begun in 1797 and completed in 1801, by which time Georgian features had been generally abandoned in favor of such Federal hallmarks as the round-topped entrance and flat-arch keystones worked into the tapered lintels, although, curiously enough, certain Adam interior details continued to be copied, as they were in this instance.

Although the names of the builder and several craftsmen who worked on the house are recorded, there is no mention of any architect, and the design is generally ascribed to the owner, George Read II (1765–1836), who personally supervised the construction.

Contributing to the imposing appearance of the mansion is the fact that it is perched on a high basement which adds low and heavily barred, marble-linteled cellar windows to the main façade, together with a central flight of wide marble steps leading up to the main-floor entrance. The beautifully proportioned archway, which spans both door and sidelights, adds equally to its charm and impressiveness. Above this arched, fanlighted doorway is a Palladian window of exactly the same width, from which projects a beautiful wrought-iron balcony. This window extends upward almost to the deep, modillioned cornice, which, like most of the exterior woodwork, is decorated with a tiny bead molding. Finally, over the eaves are two tall, pedimented dormers with round-topped windows and fluted pilasters. These in turn are surmounted at the roof peak by a captain's walk, whose exquisitely turned balusters and urn finials extend from one pair of parapeted end chimneys to the other. From this vantage point a clear view may be obtained across the river's edge, to the wide Delaware.

The post-Revolutionary Read House displays Federal details.

OPPOSITE: The drawing room combines Georgian grace with Federal charm.

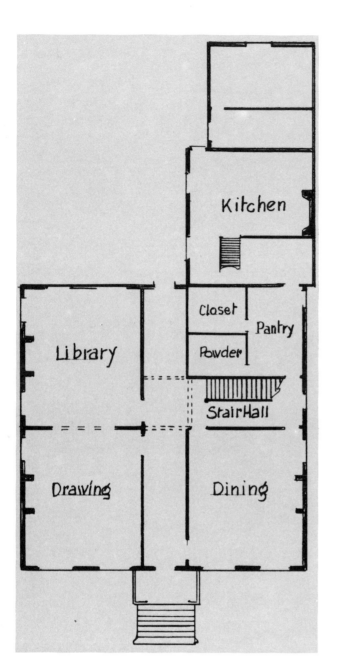

The L-shaped plan—dotted
lines indicate arches.

This principal façade fronts the almost square, two-and-a-half-story main block of the mansion, from the rear of which extends a long, one-room-wide wing, resulting in the quite deceptive L-shaped plan. In the angle thus formed is a brick-paved outdoor terrace with its summer awning alongside the garden entrance and overlooked by the two rear dormers. At the far end of this wide garden area, now devoted to flowering shrubs, spreading trees, and informal plantings, once stood the low-set mansion of the senior George Read—a signer of the Declaration of Independence—which was burned, along with a score of nearby houses, in 1824, luckily without damage to the present residence. In 1847 the area was laid out as a formal garden by the leading landscape architect of his day, Andrew Jackson Downing. Since then the luxuriant growth of trees, bushes, and shady arbors has changed its aspect, eliminating some of the formal features and clothing one entire gable with ivy to the chimney tops. Today it is a drowsily pleasant old garden whose brick walks have been trod by such eminent figures as the aging General Lafayette, George Read, the Signer, and President Taylor's Secretary of State, John Clayton, who occupied the house for a short period around 1842, and other important nineteenth-century dignitaries.

Inside the house, thanks to the huge fanlight and sidelights of the main entrance and the cream-colored walls and ceiling, the wide hall, which extends from the front portal to the rear garden door, is as light as it is spacious, particularly in the morning sun. However, since the rooms on this floor are 13 feet high, the narrower hall might seem much out of proportion were it not for two decorative arches which set off a space halfway along it where it opens into the stair hall through a third arch at right angles to the others. The two arches divide the main hall into three sections, and since these arches are lower than the ceiling, they make the hall seem much better proportioned. In addition, the heavy pilasters that support the arches break up what could have been a long, monotonous sweep of the walls, at the same time introducing decorative elements in the form of vertically fluted panels and horizontal moldings under the semicircular arches which are dotted with a line of small, blue stars from the pilaster capitals to the carved keystone, while recessed panels enliven the inner faces of the pilasters and the arch soffits.

This progression of spaces has a further advantage in logically permitting the use of three floor rugs—Shirvan and Kazak types—in place of one long runner, further reducing the apparent extent of the space without emphasizing the encroachment of the heavy pilasters into the floor area. Both of the hall's two end sections, each of which terminates in an arched and fanlighted doorway, is more than adequately furnished, that near the main entrance having a Queen Anne trifid-footed, oval drop-leaf table under a huge oval mirror and several Philadelphia Chippendale side chairs upholstered in a Burgundy-colored fabric. In contrast, the other end section features a four-paneled Queen Anne walnut settee in gold brocade and a tripod end table, all facing a handsome eighteenth-century inlaid bureau-secretary, while on the wall hangs a portrait of the Duke of Somerset by Sir Peter Lely. Incidentally, the garden door at this end has a lyre-shaped harp that strikes musical notes whenever the door opens or shuts.

The third arch, leading to the stair hall, also faces a doorway to the library, with a fine Queen Anne looking glass on either side of it, under each of which stands a Baltimore Hepplewhite chair. It is in this area between the arches that the hall lantern hangs from the ceiling, its light augmented by that of twin wall sconces in the front hall and candles in a pair of large hurricane globes on the oval table.

On entering the hall, a door immediately to the right opens into the dining room, which is one of the most charming interiors in the house, not only because of its size and proportions (which duplicate those of the drawing room across the hall) but also for the simple dignity of its exquisite furnishings and the wholly delightful panoramic wallpapers in color tones that set off the excellent fireplace mantel, which is of the Robert Wellford pattern but minus the molded figures that distinguish the originals, of which the house possesses two.

A garden view of
the ivy-covered mansion.

The old basement kitchen,
now a recreation room.

The stairs as seen through
a hallway archway.

The front hall at its junction
with the four arches.

In this room the ceiling, cornice, and dado, as well as the door and window trim, are all antique-white, a background that gives tremendous impact, not only to the mahogany doors and furniture but also to the pictorial wallpaper in natural tones, made especially for this room, which depicts various dramatic scenes associated with William Penn's landing in New Castle in 1682. These lifelike views not only show the local courthouse, but also the nearby home of George Read, Sr., which stood next to the present mansion, and the Old Tile House, built in 1687.

On the floor of this room is a rich Savonnerie rug in yellows and browns with touches of white that blend well with the wallpaper and, in the winter, with the gold damask draperies and their fringed pelmets which replace the white glass curtains used during the rest of the year. In this setting the rich glow of the antique mahogany pieces is enhanced, everywhere highlighted by the sparkle of polished silver and the gleam overhead of a lovely chandelier of Waterford crystal which, happily, is not overburdened with a plethora of pear-shaped drops so that its exquisite basic design can be observed and enjoyed.

Of all the furniture pieces in this room, the most enchanting is undoubtedly the serpentine sideboard of a Hepplewhite design, cross-banded and wearing its original eagle brasses like medals, as befits its extraordinary history. Made in Pennsylvania, it was the glory of an Early Victorian Midwest home for sixty years before its sentimental owner decided it would be happier back East in a setting for which it was obviously designed. Accordingly, it was offered to Philip Laird, owner of the Read House at that time, at a price he considered reasonable for a piece both beautiful and unique. In this same room is a twin-pedestal Phyfe dining table, the chairs being Philadelphia interpretations of a Chippendale design with carved crestings and stiles and seats upholstered in gold brocade.

On the sideboard are two six-branch Waterford-crystal candelabra, together with a tall silver coffee urn having a pineapple finial. This rests on a large silver tray, flanked by a pair of silver-mounted decanters. Nearby, centered on the dining table, is a large silver urn on a two-handled base, a beautiful gadrooned *repoussé* design with petal-like flutings, made by A. Fogelberg (*circa* 1792) and acquired from the collection of Lord Petrie.

Decorating the chimneypiece is a rosewood mantel clock, between two pieces of Export china in robin's-egg blue, together with a pair of twin-light silver candelabra whose background is a wallpaper view of the old courthouse. Other decorative—and useful—appurtenances include an eighteenth-century wine cooler on which stands a tall-necked jug in polished brass, while across the room a round-cornered drop-leaf side table accommodates candlesticks in hurricane globes and a beautifully ornamented, glass-paneled silver case in

The library with its
Wellford mantel.

which are stored a couple of decanters and a variety of wine glasses for everyday use. Actually, this room is as functional as it is tasteful, having three doors, one of which leads to the pantry, the others to the main and stair halls respectively—shortcuts to the front door, the drawing room, and the library.

At the other side of the entrance hall is the drawing room (known in more formal times as the reception room), though its doorway is near the first arch. It features an authentic Wellford mantelpiece on the projecting chimney breast between one side window and a front corner recess. This mantelpiece is beautifully decorated in low relief with figures and tiny swags modeled in French putty, but it has the usual highly intricate frieze of gouge and punch work and a molded cornice suggesting, in low relief, the triglyphs and modillions of a classic entablature, a motif repeated on the capitals of the fluted pilasters. The decorative central panel and the pilasters project an inch or more from the shelf so that there is ample space for the garniture, which consists of a pair of antique compotes and a French clock on an ebony base.

Looking into the library from
the drawing room.

Among other agreeable architectural features of this room are the heavy frieze and cornice and the rope-molding ceiling border with its quadrant-shaped corners and oblong panels in the window recesses, plus a central floral band from which depends a Waterford-crystal chandelier found in Baltimore. The moldings around the windows and doors are almost plain in comparison with the paneled window bottoms and the deep projection of the dado caps and chair rail. On the other hand, the three window tops are crowned with a wood cornice in a chevron pattern of gold and tan with a gold-leaf eagle at the center. On the floor is an antique Aubusson rug; the draperies are of red silk damask. The wallpaper, which displays a faint urn-shaped pattern in light gray over an ecru ground, duplicates that of the adjoining library.

The furniture pieces in this drawing room include an American fruitwood secretary-bookcase of 1740–1750 with its Chippendale mahogany chair, an armchair of the same age and provenance upholstered in red damask, and a fine Philadelphia Chippendale armchair, with ball-and-claw feet, upholstered in pale-green damask. By the front windows is a pair of dainty Hepplewhite chairs with oval backs and, in the third window recess, a Chippendale side chair with a carved shell on both front rail and crest. There is also a dainty pole screen in needle-point and, opposite the fireplace, a Sheraton sofa with accompanying Queen Anne tea table, plus a dish-top, birdcage table. Centered on the front window wall is an elaborate girandole with a triple candle bracket, hanging over a round-leaved Pembroke table which is also of some importance.

One extraordinary—and delightful—feature of this room is the wide, double-doored opening which connects it with the library. Over this flat arch is a semicircular fanlight reaching to the ceiling, a fantastic design of chain, swag, and loop-shaped muntins around a central half-round sunburst in wood. Inside the library this arch is keystoned, with raised soffit panels that are repeated on the underside and stiles of the opening and even in the unpainted mahogany doors.

Viewed from this doorway, the library has several exceedingly attractive features besides the solid wall of bookshelves beyond the hall doorway, including another Wellford mantel with its marble facing, a duplicate of that in the drawing room except for its garniture, and tall, elaborately carved alabaster ewers which flank an oil painting in somber colors of a dungeonlike interior, attributed to the Rembrandt school. Other features similar to those of the adjoining room are the ceiling decoration, trim design, and wallpaper, though this room has a window on each side of the chimney breast instead of only one. In addition, both of the end windows are of the gib type whose bottom panel can be opened to permit an exit to the rear terrace. All these windows have floor-length chintz draperies in brown and green on a beige ground, the small pattern tying in nicely with that of the large Khurasan carpet.

Between the end windows hangs a tall Queen Anne looking

The dining room with the wallpaper made especially for it. Scenes of New Castle.

glass over an eighteenth-century, American three-drawer table in walnut with pad feet. To one side of this group is a Chippendale straight-legged wing chair in pale-green upholstery, and a similar one under the right-hand window.

Alongside the former is an oval inlaid Pembroke table burdened with a reading lamp that balances a floor lamp in the opposite corner. In front of the bookshelves is a particularly fine Chippendale armchair with scroll arms, matching the carved side chair just inside the drawing room. Conveniently near it is a pad-footed Queen Anne tea table which can readily be drawn in front of the camelback style, roll-armed sofa, upholstered in a green damask, which has a place of honor in front of the hearth. Though the bookshelf itself is of little note, its top does serve to display a number of interesting items, such as an English shelf clock of about 1740 and a variety of Chinese Export porcelains.

Directly above this room, on the second floor, is a particularly charming sitting room from whose landing doorway can be seen not only the stair to the third floor with its Oriental-looking hanging lantern, but also the great Palladian window which lights both the lower stair hall and the upper landing. This window is even larger than the one over the front door, and though apparently no part of the total design, it adds a charming detail to that side of the house overlooking an adjoining alley.

The sitting room itself luckily has plenty of daylight illumination through its four windows and the almost ever-open door that connects it to the owner's bedroom. Here, too, is another copy of the Wellford mantel minus the molded figures of the original. More notable is the lovely little block-front kneehole desk serving as a dressing table, with its gleaming willow-pattern escutcheons and lifting handles. This is surmounted by a smooth-framed Queen Anne mirror, and next to it is a comparatively tall Queen Anne chair by William

Among the American-made pieces a blockfront, bonnet-topped highboy.

137

View from the library into
the drawing room.

Savery, with its dignified scroll splat and a balloon seat in a soft pink dotted with white roses. Occupying the center of the room is a mahogany-framed Sheraton settee in the palest of green silk brocades, accompanied by a straight-legged armchair in crimson damask as counterpoint.

The white ceiling is beautifully decorated in the manner of Robert Adam, its huge oval garland centered on a radial flower-and-fern cluster reminiscent of the garden room at Osterly. A gold-framed family portrait above the white mantel stands out against the pink-tinted wall with its bold splashes of yellow and green, the latter color picked up by the porcelain vases which, along with some entrancing miniature figures in chalkware, form the mantel's garniture.

The foregoing speaks for the rest of the rooms which follow the same pattern of genteel luxury in their adaptation to present-day needs and tastes at the hands of the mansion's owner, Mrs. Philip D. Laird, who has also succeeded in preserving its early-nineteenth-century character and the air of vitality with which it must have been endowed by the man who designed and built it 170 years ago.

OPPOSITE: One of the house's most charming interiors—the dining room.

THE THOMAS-JENCKS-GLADDING HOUSE

An Astonishing Mid-Nineteenth-Century City Survival

One of the finest mid-nineteenth-century urban residences in the United States is the downtown Baltimore mansion known as the Thomas-Jencks-Gladding House after the three owners who have successively occupied it during the 120 years of its existence. The last-named is Harry Lee Gladding, the present owner, who endowed it with a new lease on life after 10 years of neglect and consequent deterioration when he rehabilitated it in 1963–1965 to serve as his private residence.

Designed and built in 1849–1850 by architects Niernsee and Neilson for Dr. John Hanson Thomas, this sturdy, three-story brick structure, whose exterior has been little changed over the years, is made even more imposing by a raised basement, the height being further emphasized by tall French windows with classic bracketed pediments on the first and second floors, the third-floor windows being of the stub type, tight under a stepped, brick frieze. Above this point the overhanging cornice of the almost flat roof is upheld by closely spaced modillions and ornamented with a row of palmette-style antefixes of cast iron.

An interesting feature of the main and end façades is the slight recessing of the outer walls to break the monotony not only of the flat surfaces but also of the cornice, the stone stringcourse, and the water table, while adding to its appearance of solidity. On the main façade, the entrance portico is of the same width as this recess, rising to the level of the stringcourse over a paneled base of cut stone which encloses the marble steps. On this base stand four fluted columns of the Composite order, supporting a plain entablature. These are reinforced by two more columns flanking the entrance door and a pair of plain pilasters that finish off the brick corners of the recess.

The importance of this center section is further emphasized by making the second- and third-story windows above it three panes wide by inserting twin stone mullions. All the tall windows are further improved in appearance by decorative,

The Thomas-Jencks-Gladding House today.

OPPOSITE: An exquisite setting for formal dining.

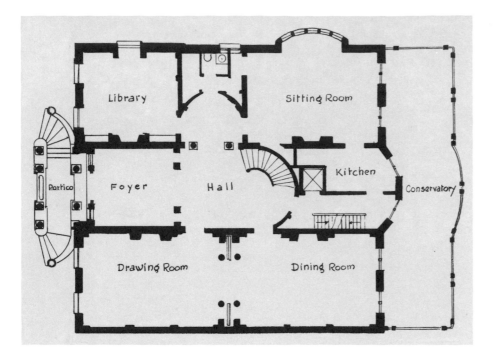

A roomy town-house layout.

Beyond the foyer is the rich central hall.

cast-iron safety grilles, except for the two on each side of the portico, which are tied together by balconies with rails of the same height, while bushes beneath them hide the basement windows.

Originally the house had been designed to accommodate the prevailing furniture style of the declining Greek Revival period, and it was therefore well suited to the Gothic Revival pieces which the Thomases favored, though they later introduced the somewhat newer Elizabethan Revival style, with its Charles II corkscrew turnings, with which the original architectural and decorative features presumably did not harmonize.

In 1892 the house was sold to Francis Mankin Jencks, and shortly thereafter some interior and exterior changes were made by the brother of Mrs. Jencks, the noted architect Charles A. Platt, who substituted Italianate details to create a sympathetic atmosphere for the furniture he had bought for the house in Italy. At the same time he widened the staircase, endowing it with a more graceful curve, and reduced the height of the library ceiling. He also substituted a large bow window for the plain one in the former dining room, which is now the sitting room, and when the property was taken over by Mr. Gladding in 1963, along with some of the Jencks' furniture, the whole house was in a dilapidated condition after ten years of neglect, although the exterior had more recently been stripped of its shabby coat of paint, and the brickwork repointed, to restore, as far as possible, its original aristocratic air. The whole interior was therefore repaired and refinished, including the exquisite plasterwork, some of it still Greek Revival in mood, though Mr. Platt had replaced part of it with Italian Renaissance designs, which luckily provided a more acceptable background for the preponderance of Louis XV and Louis XVI pieces with which it was eventually to be furnished.

Today this difference can be observed immediately upon entering the mansion through the shallow portico illuminated by a lantern from the Parisian Hôtel Bristol. Once inside the palatial foyer, the urge to explore farther is almost irresistible. The white-marble foyer floor, sprinkled with jet-black cabochons, stops short of the tall, fluted columns beyond which this fabulous interior really begins. Nevertheless, in this short space the few pieces of furniture are indicative of what is to come. To one side is an incredibly rich Louis XIV console table, its marble top garnished with a Dresden bowl and Louis XV candelabra. Above them hangs a Rembrandt Peale copy of a painting by Rubens in the Pitti Palace. Alongside the table are two decorative Louis XVI gilt chairs and a large blue-and-white Chinese vase. Against the opposite wall is a similar table with Empire-style candelabra beneath a portrait of Charles A. Platt by Kenneth R. Cranford, done in Paris.

Beyond these the black-marble border of the flooring changes suddenly to mosaic-pattern parquet, on which is centered a ruby-colored Sarouk rug patterned in rose and blue. The walls are of a pale beige, the trim white, and to one side is the

A detailed view of the fantastic coffered dome.

A circular stair winds its way to the upper hallway.

Foyer furnishings include a Louis XIV console and a Peale copy of a Rubens painting.

pale-beige painted stairway sweeping upward in a semicircle to the first landing. Both the soffit of the stairway and the L-shaped hall ceiling are coffered, and from the latter hangs a sparkling Louis XVI crystal and bronze chandelier.

On the main floor, by the stair's curtail step, is another huge Chinese temple jar, and in the rear wall which follows the curve of the stair is a Roman-arched doorway to the kitchen and conservatory, flanked by a transitional Louis XV marquetry commode and an Italian marble bust of Hadrian on a black-marble pedestal. Between the two wide doors to the dining and drawing rooms is arranged the principal group, consisting of a round Italian mirror in an extra-heavy, Rococo-style gilded frame and, below it, a Louis XV marquetry commode whose white-marble top is adorned with a pair of candle lusters and a colorful Venetian-glass goblet. Alongside stand two more of the fragile-looking Louis XVI turned and gilded chairs, one of them in the faux-bamboo pattern.

In each corner by the columns is an upholstered tall-back dining chair in crimson brocade with bronze *doré* crook-shaped finials belonging to the set brought from Italy by Charles Platt. But, standing among these treasures, once the eye falls on that marvelous stairway, all else is forgotten. From the center of this floor one can see up through the circular stairwell to the fantastic coffered dome with its oval skylight of Tiffany stained glass and the elaborate stucco decoration.

On the second floor the curving wall is pilastered and

The dining room features a Louis XVI marble-topped commode.

paneled. Above that is the dome's interior enlivened with huge fingerlike panels and hundreds of star-petaled flowers in high-relief stucco, the whole thing as enchanting in detail as it is in total effect.

Closer at hand are the doors to the drawing room and dining room respectively, and once inside either it can be observed that the two interiors, which are separated only by paneled sliding doors, duplicate each other insofar as the walls and ceilings, the fireplaces, chandeliers, rugs, draperies, and the unframed, room-high pier glasses are concerned.

The white stuccowork in these rooms is superb, emphasized by the three beige tints used on the flat surfaces to stress the ceiling-panel borders and the background of the chandelier rosettes, as well as the friezes, the eared door surrounds, and the heavy pilasters that break the walls into panels and so minimize the need for other wall decoration. The chandeliers, of Baccarat crystal and ormolu, were brought from Paris by Dr. Thomas about the time the house was built.

The decorative qualities of the stuccoed ceilings are at least matched by the 1840 English-made Savonnerie rugs with their oversized floral pattern in tones of beige, rose, and dark blue. The draperies, which have deep swags and unbalanced cascades, are of beige cotton brocade, slightly darker than the walls and bordered with tiny light-beige tassels. Of equal importance from a decorative standpoint are the heavily bracketed wooden mantelpieces with their stucco decoration and wide facings of white marble. On the other hand, the dining room mantel has three decorative panels between its four brackets and is painted to match the ceiling, whereas the drawing room mantel is all white.

Major features of the dining room are the tall-back chairs like those in the hall, disposed around a rosewood banquet table which, when in use, is covered by an Italian cut-work cloth. On the other hand, the most elegant piece of furniture in this room is the Louis XVI marble-topped mahogany commode, elaborately decorated in ormolu tracery and gilt-bronze plaques, the whole thing copied from one made in 1787 and now reposing in the Louvre. Its drop front discloses shallow drawers, possibly intended for the storage of silverware. On it are twin Louis XV candelabra in gilt bronze, with an early nineteenth-century English family portrait on the wall above them.

The drawing room is possibly even more beautiful, thanks to its formality, its outstanding piece a magnificent Louis XIV marquetry commode in bronze *doré* with four cast-bronze figures in high relief adorning the stiles. This supports a white-marble bust of Marie Antoinette and two tall Sèvres candelabra, gold-mounted in the characteristic blue, flanking a self-portrait of Vigée-LeBrun (1755–1842), copied from the original in the Uffizi Galleries of Florence. Facing this across the room is the fireplace with its *garniture de cheminée*, which consists of a similar Sèvres vase converted to a clock and

In the northeast guest room stands this English chest beneath a Rococo mirror.

At one end of the drawing room is this Louis XV marquetry *bureau plat*.

The drawing room's outstanding piece is the Louis XIV commode.

A tall Venetian mantel relieves the library's somber air.

mounted on a footed bronze base with *cherubini,* accompanied by two small porcelain urns in the same deep blue, white, and gold.

The seated pieces in this room are a mixture of Louis XV and Louis XVI *bergères,* chaises, a marquise, and a lovely *canapé à corbeille.* Behind this sofa is a splendid Louis XV marquetry *bureau plat* with ormolu masks, chutes, and sabots. On it rests a fabulous blue and gold bowl with painted scenic and floral panels mounted in gilt bronze—a riot of fruit and flowers in the lavish Rococo style.

In sharp contrast to this gay and lively room is the library with its background of dark woodwork covering the ceiling and walls and introducing a rich but somber air. Happily, considerable relief is afforded by the tall Venetian mantel in a gray and white marble with its finely carved frieze and cornice and vase-shaped engaged columns. This huge mantel is adapted to a small fireplace opening by the use of wide and deep white-marble facings, relieved only by a narrow ledge beneath the mantelshelf on which decorative objects can be placed to distract the eye.

All of the woodwork in this room is of mahogany, including the exposed (actually superimposed, since the original ceiling is *in situ*) ceiling timbers and the carved beams, ornamented with carved modillions, dividing the joists into deep rectangles inside which the oblong patches of exposed ceiling are covered

The north bedroom features this bed *à la polonaise.*

Mrs. Gladding's sitting room.

In the central hall this round Italian
mirror is grouped with
candle lusters and a Venetian goblet.

with gold-flecked paper within a carved and parcel-gilded frame. The walls, too, are divided into panels—this time large vertical ones—by mahogany pilasters extending upward from the dark-paneled dado to the mahogany frieze, their monotony broken, here and there, by floor-length windows and the glass-doored bookcases alongside the projecting chimney breast.

These pilasters are not only blind-carved in a fretwork pattern, but their tops are formed into cartouches with cabochon centers, while spaced at intervals down their length are projecting heads of notabilities carved in the round. Between the pilasters the walls are covered with gold-and-aqua silk brocade in a pattern also used for the draperies. Artificial illumination is provided by a Louis XVI-style crystal chandelier, augmented by twin-light wall sconces of bronze in a torch design.

Of equal interest with this unusual parquet-floored interior is the mixture of Louis XV, Napoleonic, and Victorian furniture pieces which somehow seem to live happily together, on or around the pale-gold Chinese rug. Louis XV is also represented by a distinguished *bureau plat,* a leather-topped marquetry piece whose ormolu mounts include chutes in the form of a completely modeled female head. In contrast with this, though not lacking in curves, are the sturdy French Empire gilded armchairs with stout winged griffins and lion paws, the seats and backs covered in gold satin, though the simple, armless chair has neither heads nor wings, merely the legs and paws and a beautifully carved frame. Sharing room space with these are French chairs of the Victorian period with Louis XVI legs, seats almost circular, and rounded backs with spindle supports, padded on top to act as both back- and armrests.

There are many more rooms in this interesting and delightful mansion, and some lovely mantels in French and Italian styles, with walls on which beige brocade replaces oak, or which display a collection of Piranesi prints on ecru-tinted plaster. The furniture ranges from Biedermeier pieces to an English field bed—all quite accidentally displaying an eclectic taste that has resulted in the creation of a wonderfully interesting home out of a pseudoclassical Victorian mansion that but for one man's vision would most probably have encountered an ignominious fate.

Across the rear stretches the conservatory.

Mr. Gladding's austere bedroom.

149

Old Westbury, New York

WESTBURY HOUSE

An Eighteenth-Century English Manor on Long Island

A close-up of the residence's main façade.

Probably the most striking feature of Westbury House is the air of serenity that pervades it, both inside and out—an ivy-clad mansion of limestone-trimmed, cherry-red brick, nestling among ancient trees and lovely gardens as though it had been a part of the landscape for centuries past. Such an atmosphere no doubt is due in some degree to the fact that this charming structure is a historically accurate representation of an eighteenth-century English manor house, furnished largely with period pieces, plus later additions and minor changes that suggest the rural dwelling grown gracefully old. Actually the house was designed and built by a London architect, George Crawley (1864–1926), as a country residence for the well-known financier-sportsman John S. Phipps, in 1906.

Approaching the mansion via the long, curving driveway between tree-dotted lawns, the visitor's first view is of its gold-tinted roof of Ketton ironstone from tiny Rutland in the heart of England. Beyond this is the main façade whose wide entrance door, with its tall, small-paned sidelights, is sheltered beneath a limestone porch the height of the stringcourse, with three ceiling-high windows on either side, so that its roof forms a balcony accessible from a fanlighted French door which opens from the second-floor hall. This door, in turn, is flanked by oval *œil-de-bœuf* windows set on their vertical axes. These, along with six other second-floor windows, are tucked neatly beneath the roof's modillioned cornice. The façade is completed by a huge, central arched dormer—which adds a Queen Anne touch—incorporating a Venetian window, with two regular pedimented dormers on either side set into the steep mansard roof. Centered on the flat top of this roof is a massive flat-topped chimney grouping pierced by an archway. This is balanced by tall stone chimney stacks paneled in brick (a seventeenth-century conceit) above the end walls of the house.

The rear façade is much more spectacular than this, the lawn level being 15 feet lower, allowing for twin flights of steps beyond the stone-balustraded rear terrace, whose high wall

is almost hidden under cascades of wisteria, revealing, at intervals, limestone niches containing antique marble busts. Fronting the flight of steps at lawn level are cone-shaped topiary yews, facing on the seemingly endless *allée* of lindens with an orchard on one side and the rose garden on the other.

Above all this rises the south façade, whose projecting end pavilions are crowned with deeply hooded, round-topped casements supported by carved consoles, the setback main roof harboring five pedimented dormers similar to those on the other three sides of the steep roof. Beneath this cornice the house wall has a window below each of the dormers, the central one being a slightly larger twin-casement style with an elaborate scroll pediment and heavily carved stiles in the Baroque manner. At terrace level is a sturdy stone balustrade, beyond which a short, flagged path leads to a flight of stone steps, guarded by eighteenth-century English sphinxes whose feminine features are more delightful than horrendous. As in the case of the main (north) façade, the central arched chimney grouping adds a final fantastic touch.

At each end of the house is a single-story addition. That against the west wall consists of a glassed-in porch designed by George Crawley, which was added in 1924. This also is set off by pairs of coupled Ionic columns that replace the usual wall and serve to support the balustraded roof, together with the steel tracks of the vertically sliding glass walls which can be lowered into the floor.

At the other end of the house a quite different style of structure with the same brick walls and stone-clad roof as the main building contains a huge kitchen, sculleries, store rooms,

One of the twin stairways
beneath cascades of wisteria.

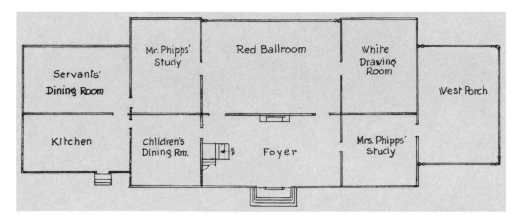

The layout of the house's
main floor.

Mr. Phipps' study and his
favorite chair.

and a dining room and bedrooms for the servants, all designed
by the Philadelphia architect Horace Trumbauer (1869–1938),
in 1911.

With such an attractive exterior it is no surprise to find the
interiors equally full of character, thanks to the atmosphere
created by their architectural details, each room being dis-
tinctively different. In the front hall, for example, these features
are sufficiently prominent and decorative to eliminate the
necessity for much in the way of furniture. The floor is of
alternating gray and white marble slabs, laid diagonally, their
texture so well in keeping with the limestone walls and the
huge carved fireplace mantel. This mantel, which was designed
by Derwent Wood (1871–1926) of London, has exquisitely
carved consoles let into the stiles, a central panel depicting the
race of Atalanta's suitors, after a drawing by Michelangelo, and,
reclining on the two halves of a semicircular broken pediment,
a pair of life-sized *amorini* representing Time and Love, with
a Wedgwood basalt bust of Milton on a pedestal between them.
At the top of the overmantel a cartouche and swags with drops
carry the eye up to the paneled ceiling with its great fresco
depicting fleecy clouds in a sunny sky, all enframed by a
flower-adorned stone balustrade, the work of A. D. Carse.

At one end of the hall is the principal staircase, terminated
at the bottom by a pair of tall, attached Corinthian columns,
each fronting a square pillar and topped by a segmental arch
with reclining figures, tying together a room-wide modillioned
entablature, supported by two pairs of similar columns, spaced
to give access to deep alcoves alongside the staircase. The most
important feature, however, is the carved and pierced paneling,
between a series of heavy consoles—actually forming an oak
screen—in the space between the entablature and the coffered
ceiling.

Opposite the fireplace is the huge entrance door with its
semicircular fanlight and side windows. From the center of the
great ceiling painting hangs a heavy bronze chandelier—
probably an English adaptation of a Louis XV style—that is
perfectly suited to the carved woodwork and massive fireplace.

The few furniture pieces, too, have that amazing feeling of
delicate solidity, graceful but strong, that adds to the attrac-
tiveness of the interior without detracting from the feel of airy

space. Under the chandelier is a late-seventeenth-century William and Mary library table, its apron inlaid with seaweed marquetry, its polished top reflecting the shape of a large and lovely Imari bowl seated upon it. On either side of the fireplace is a Daniel Marot-style seventeenth-century, English high-back walnut chair, and beyond them, an Irish Chippendale side table of about 1775, with a *verde antico* marble top, and an equally old Georgian eagle console whose top is marbleized. Across the room from the stairs is a late seventeenth-century tall case clock decorated in marquetry, its movement by Thomas Trigg, and a Restoration period walnut desk believed to be that on which Oliver Cromwell signed the death warrant of Charles I.

Although impressive, this hall by no means reflects the beauty and comfort of the drawing room, which is exquisite in every detail. This gracious white-walled interior, with its paneled and dado-molded walls, has tall windows draped in

Despite little furniture this hall is full of character.

The second-floor landing
displays several unusual antiques.

Richelieu peers out from this
portrait in Mrs. Phipps' study.

white and yellow damask under delicate Chippendale cornices and shaped, cascade-type valances. The fireplace is of carved Carrara marble, with consoles plus a paneled frieze centered on a feminine mask in high relief. The overmantel is in the form of an eared panel in which is set an oil painting by Sir Joshua Reynolds. This whole chimney breast is emphasized by attached Corinthian columns and pilasters supporting a complete entablature, the heavy modillioned cornice alone being carried right around the room. From the plain ceiling hang two Waterford-crystal chandeliers, which are reflected in a huge gilded Chinese Chippendale looking glass hanging between two of the windows. Much of the room's atmosphere, however, is created—and the furniture groupings unified—by the 1760 English pile carpet whose brown checkered ground is enlivened by bouquets and sprays of flowers in their natural gay colors.

Among the English antiques are a walnut pedestal-style desk; a tall case clock in the Gothic manner with a movement by Henry Rigby of Liverpool; a 1760 Chippendale mahogany stool covered in gros-point and petit-point floral fabric; a large Chippendale pole screen in tapestry, variety being supplied by neoclassical satinwood pieces, including a pair of console cabinets and several small tables. Interspersed among these antiquities are luxurious modern sofas and deep chairs which add tremendously to the air of luxurious comfort while providing great splashes of color and pattern that satisfy the aesthetic sense. Of primary interest, however, are the paintings by such notabilities as John Constable (1776–1837), John Crome (1768–1821), Sir Joshua Reynolds (1723–1792), Richard Parkes Bonington (1801–1828), Angelica Kauffmann (1741–1807), George Morland (1763–1804), and several family portraits by American and English artists of note, all of which add a certain richness to the atmosphere.

Quite different in feeling is the oak-paneled dining room, which not only possesses two original Grinling Gibbons (1648–1720) carved pendants but also an extraordinarily fine white-marble mantelpiece having a sculptured term at each end and a contrasting molded facing in *verde antico*. Above the overmantel panels a carved frieze in high relief continues the fruit and flower theme of the pendants right around the room. Similar narrow decorative panels in natural wood border the ceiling, each of them interrupted at the center by a large scrolled carving in the round which extends into the coved plaster cornice and out toward the huge ceiling fresco which depicts a cloudy sky. All the overdoors, beneath their segmental pediments, and the door stiles are carved, as is the molding of the panels, dado, and baseboard, while the ceiling is trimmed with stucco detail.

All this has resulted in an extraordinarily handsome setting, accented by silver plaque-style wall sconces, a silver-plated *repoussé* chandelier, and blue-velvet draperies. The collection of paintings here is equally fabulous, that over the mantel

The dining room displays Gibbons' carving and portraits by famous artists.

being a Gainsborough portrait of the Duke of Clarence, the rest including works by Raeburn, Sargent, Reynolds, and the Irishman Stephen Slaughter. Notable furniture pieces comprise a pair of carved and gilded eighteenth-century side tables with scagliola tops; a somewhat earlier pier glass; and an inlaid mahogany Hepplewhite sideboard, with a console table to match, all of which must certainly be exceeded in interest by a complete set of Irish walnut, fan-back, green-damask-covered chairs used with the massive, carved dining table of a less distinguished pedigree. The carpet is English Worcester in green.

Off the foyer is Mrs. Phipps' study, which is as elaborate in detail as any room in the house. Paneled in pale green and white, it has two walls devoted solely to bookshelves above sill-height cabinets. Above the plain frieze is a cornice molding in an intricate acanthus-leaf pattern. Though this is quite decorative, it is far less bold and vigorous than the three-dimensional circular band of stucco molding that sets off the ceiling's coved center, from which hangs a large gilt-bronze and crystal chandelier.

Equally impressive is the unusually detailed fireplace mantel-piece, set between two windows, whose pedimented overmantel touches the cornice and barely leaves room on the chimney breast for a pair of gilt-bronze sconces. The mantelpiece itself is of white marble, its flower- and fruit-adorned stiles reinforced by rose-bedecked scrolls topped by male busts reaching up to the architrave, the latter having a central panel carved with *amorini* and a lamb in low relief. The overmantel is even more ornate, its pediment supporting two laurel-crowned masks

The Chippendale guest room.

The west porch looks out over the terrace.

The small but delightful Adam bedroom.

from whose beards hang flowering vines alongside the scroll baskets which, along with the fret-carved base, enframe a portrait of Cardinal Richelieu.

In addition to two modern upholstered pieces, slipcovered in a floral print on a pale-green ground, this small room finds space for a Chippendale mahogany breakfront bookcase, four Hepplewhite shield-back armchairs, and a fretwork-galleried, birdcage-type tripod table plus a Chippendale fireside stool. There is also a lovely gilt-bronze mantel clock by Benjamin Gray, clockmaker to George II. The draperies are of green and white damask.

The study used by Mr. Phipps is, as might be expected, quite masculine in taste, even though there are many purely decorative features in the Robert Adam or James Wyatt manner. This room has the same kind of coved ceiling, but the whole flat area outside the coved area's periphery is lavishly decorated in stucco and even takes part of the room's cornice, whose only decoration is a row of dentils where it joins the green-painted paneled walls. Both the chimney breast and the mantelshelf are encased in wood, though the fireplace opening has a heavy marble facing. Applied decorations in the form of swags, drops, and posies, all painted white, embellish the woodwork both above and below the mantelshelf. A decorative panel centered in the overmantel contains a Flemish still-life painting, the space between the mantelshelf and the stone molding being occupied by a horizontal mirror, flanked by the stucco floral clusters.

This room is floored in mosaic parquet squares, rightly innocent of rugs as a background for the antique pieces. Quite unusual is the old-time group of two swivel-seated Burgo-

meister chairs—one in walnut, the other in mahogany—of the early 1700's, and a leather-topped Hepplewhite library table of half a century later, whose three-lobed pedestals form capacious cupboards.

Another interior that concentrates several fascinating items in a small space is the second-floor landing, which actually is a room. The most important of these pieces is a tremendous bun-footed English wardrobe, veneered all over in delicate marquetry, details of which date it as having been made shortly after 1684. Of almost equal antiquarian interest is the white-marble mantelpiece, which is a perfect duplicate of one designed by Sir Henry Cheere (1703–1781) for Eltham Lodge in Kent. This has inset panels of colored and figured marble of the *fleur de pêche* type.

Another unusual piece in an American home is the oval Hepplewhite writing table of inlaid and beautifully grained mahogany with a leather top. Above it hangs a splendid Waterford chandelier, a final touch that helps convert a pass-through hall, with its Charles I chairs, into a place to linger.

In all these examples of Westbury interiors, which are characteristic of the rest, it will have been observed that tradition has been combined with originality, and charm with comfort in the modern sense.

Today the manor house and its well-kept grounds remain as invitingly fresh as they did when the family was in residence, thanks to the devotion and the expertise of the Old Westbury Gardens Foundation, whose business it is to preserve the interest and authenticity of the manor and its 100 acres of statuary-dotted *allées,* walks, and gardens for the benefit of all.

In the drawing room is an overmantel painting by Reynolds.

Details of the Rococo mantel in Mr. Phipps' dressing room.

Athens, Georgia

THE PRESIDENT'S HOUSE

Classical Design for Modern Living

The house appears enframed
in masses of foliage.

One of the South's loveliest mansions is the President's House of the University of Georgia. This palatial domicile, so correctly classical, was erected in 1855 by a Virginian named John Thomas Grant and acquired by the university, in 1949, from the estate of its latest owner, Mrs. W. F. Bradshaw. In that same year it was furnished in a manner befitting its perfection of design, and today it stands in its nine acres of lawns and formal gardens as a superb example of the Greek Revival style at its best.

Looked at from the front gate, the mansion appears enframed in masses of crowding foliage—a leafy bower of clipped tree box, cherry laurel, magnolia, crepe myrtle, jasmine, and syringa, and other shrubs and trees that thrive in this beneficent climate. As one approaches along the wide, brick-paved walk, this flat-topped, white-painted gem comes into full view, presenting a delightful picture with its eight tall, fluted Corinthian columns spread across the raised veranda under the widely overhanging eaves of an invisible roof. The decorative quality of the columns is enhanced by the smooth sweep of an entablature—quite plain in comparison but exquisite in proportion—that emphasizes the horizontal.

This Corinthian peristyle not only extends across the front but also along the sides of the house, halfway to the back where outward jogs in the building set a limit to the side porticoes. At the rear the building center is recessed to accommodate a short portico with four Doric columns, beyond which two single-story structures—one housing a kitchen and breakfast room, the other a guest bedroom—project. Here the rear staircase descends into a paved, rectangular space between the extensions. At the second-floor level there is an equally short balcony behind the columns, emphasizing its architectural grace.

Opening off the front portico, the main entrance to the mansion, with its Corinthian pilasters, is topped by a similar door at second-story height (minus the Corinthian accent) which opens onto a balustraded balcony. At first-floor level, the two pairs of windows flanking the doorway extend upward from the porch floor to almost ceiling height. In contrast, the upper ones, which have thin, bracketed sills, are well up from the floor, but both main- and second-floor windows do have the same style of heavy, flat pediments, imitating the roof arrangement minus the lovely capitals.

Since there is a wide basement entrance at ground level in the center of this façade, with steps leading down from the brick walk, the center section of the porch floor is extended forward over it to accommodate two iron-railed stairs that sweep down to ground level in wide, converging arcs. Now partly overgrown with creeper, these curving steps actually add to the imposing appearance of the main entrance and the upstairs balcony which share the space between the wide-set center columns.

At the rear, the house is only slightly less impressive, its wide lawn beautifully landscaped, its stairs flanked by tall bushes, and the brick walk leading in both directions around the neatest of small lawns. This neat grass oblong lies there like an exquisite carpet, stretching out between six square, stone-capped, low brickwork islands full of box and low bushes, toward the great grass plot beyond, all hemmed in by towering trees. On either side of this small grouping of lawn and planters is a single-story building—one an arcaded cottage; the other

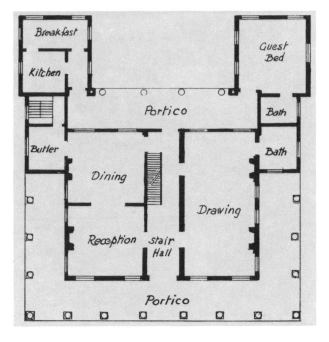

Plan of the President's House.

The mansion from the front walk.

One of the Corinthian columns
and the classical eaves.

A view across the portico with
its central balcony.

a rather less-neat old carriage house, both hemmed in by low box hedges.

Close by the cottage gable, 4-foot-high brick walls forming open-latticed panels between brick pillars enclose an old-time rose garden. In this wall a white, square-picketed gate opens onto a large, brick-paved area whose four L-shaped flower beds surround a sundial. Borders of standard roses and rows of box line the walls, while here and there a long, white-painted, wrought-iron, five-back bench, all loops and curves, begs the wanderer to rest—and observe.

Architecturally, the three major main-floor rooms of the mansion are every bit as interesting as the exterior, though quite unalike one another. The drawing room, as might be expected, is by far the most elaborate, largely because of its remarkable frieze and cornice, which are almost Baroque in treatment, and the bands of foliage alternating with cartouches and fruit clusters. This is painted a dead white. The two large and elaborate Regency chandeliers of 1810–1820 are draped with clusters of sparkling thumbprint crystals, together with others in the shape of stylized honeysuckle blossoms—a favorite Adam theme. These fixtures, which were original to the house, hang from intricately carved ceiling roses which are even greater in diameter.

With its seven floor-length windows, three doors, and two fireplaces, this 20 by 40-foot room has no space for a dado, so crowded are the walls. But it does have a huge pier glass, and both fireplaces are of a Victorian arched style in white marble, their overmantels occupied by large oil paintings of Roman ruins in the Pannini style, which are admirable in execution, though somber in effect. Both walls and ceiling are of a peculiar bluish-green color, punctuated by taffeta draperies in a slightly darker shade, and topped by heavy, fringed, draped pelmets of the same material in gold. On the floor is a velvety, modern, room-sized rug in sage green. The total effect is quite striking and perfectly suited to the mahogany and upholstered pieces spaced around the room in such a manner that there is no feeling of clutter.

A welcome color relief is provided by a pair of pink-and-white-striped Hepplewhite sofas in satin brocade, one at each end of the room. There are also touches of gold here, besides that of the pier-glass frame; two of the delicate Louis XV-style *fauteuils* in mahogany, with cabriole legs and scroll feet, are covered in gold fabric, while the pad-footed pair of tall-back armchairs use a white fabric with gold and brown patterning. Both mantels have garnitures in ebony-trimmed gold: one consisting of a clock and two vases on pedestals; the other a pair of crystal-hung candelabra. Between the two windows separating the fireplaces hangs a small antique picture almost entirely gold. Of the upholstered pieces one is yellow, the other beige. A round mahogany pedestal table with drawers and a small occasional table with round, ebony legs add a Regency touch that emphasizes the eclectic style of this impressively beautiful interior, which strikes a happy medium between simple splendor and pretentiousness.

Even more pleasing, however, is the somewhat smaller and less exuberant dining room which, in spite of its formality, is both charming and comfortable, while representing the acme of good taste. The setting is exquisite, its ceiling, trim, and dado—the latter composed of plain moldings—all in white, emphasizing the three-dimensional qualities of the reproduction Zuber wallpaper, which depicts a rugged landscape with ruins amid tall trees, boar hunters, and riders on horseback. This shows so much sky and distant horizon that the room seems much larger and airier than it really is. Augmenting that feeling of spaciousness is the wide, flat archway at the rear of the room, opening into the reception room which affords a still farther view out of its front windows. This dining room also opens into a butler's pantry at the rear of the stair hall which leads to the rear portico.

The dining room's projecting chimney breast accommodates a fireplace in a mottled-brown marble, over which is hung a large, Federal-style bull's-eye mirror, flanked by crystal candelabra. The *pièce de résistance,* however, is the English crystal chandelier, made by Perry in 1810 or thereabouts, with its pear-shaped hanging drops. Centered beneath this is the mahogany extensible dining table, a copy of an Empire piece, paired with a 1780 Hepplewhite-style, serpentine-fronted sideboard with oval inlays, while to one side a large butler's tray on its stand acts as a serving table. Quite often this is adorned with a sterling silver set of beakers and a water jug on a gadroon-bordered tray. The dining chairs are an unusually comfortable type with fabric-covered, rounded backs and upholstered seats in a cotton floral brocade, whose colors harmonize well both with the lavender-tinted rug and the draperies of apricot taffeta with a Roman border.

In the reception room the detailing is even simpler since there is neither wallpaper nor dado, though there is still plenty of character. The walls are a soft gray-green, the ceiling and trim white, and the rug an inconspicuous taupe. The draperies, however, are lively enough, being of linen with a floral pattern in blues and grays. These colors are picked up in the upholstered pieces which contrast striped and diaper-patterned fabrics with plain materials in monotone. The accessories are confined to several handsome lampstands or occasional tables, both galleried and plain. The fireplace is a duplicate of that in the dining room, a streaky brown-marble inset with a fire frame and grate and fronted by a bronze-decorated screen of a Louis XV style with a wreath and torches and a central element of bagpipes, sword, and leafy sprays.

The mantelpiece garniture consists of a Louis XVI carved, marble and bronze clock with an open pendulum and a pair of yellow, *tôle* urns. Above these is a gold-framed portrait. The colorful window draperies in a floral pattern enframe both the chimney breast and mantel.

The rear portico with its
two-story fluted Doric columns.

The rear portico from the garden.

The dining room and its pictorial panels.

Elsewhere in this room, which opens into the hall, are two other interesting groups that give it character. One of these is a window arrangement comprising a mahogany, Regency-style stand with a single drawer and a tall baluster-type gallery, on which stands a Chinese-looking lamp utilizing a square porcelain vase in blue on white, with an angular, footed teak base, all set off by the draperies, which hang outside the paneled window reveals, and a mahogany chair in brass-nailed purple leather. The third group is centered on a neat but handsome Sheraton chest or serving table—an antique, inlaid tea caddy, a Lowestoft, medallion-decorated, and quite small jardiniere in white and gold, plus a small piece of Export china. On the wall above is a carved and mirrored Victorian bracket, with a tiny blanc-de-chine bust of one of the Caesars on its shelf.

Just outside the drawing-room door in the rear hall is an elaborately carved Victorian triple mirror in rosewood, with a very low base and eight shelves on brackets in *étagère* form. On the other side of the door is a small Empire console with columns and recessed drawers. On it stands a tiny metal bush with a bronze and scaglioli base. This is given importance by an elongated twin-candle sconce, apparently Italian, of carved wood and black iron on the wall behind it. Next comes a Late Empire, animal-footed sofa in green-and-white-striped silk damask, under a beautifully framed painting, *Boating on the Seine,* by Robert Spencer, from the Alfred Holbrook Collection of the Georgia Museum of Art.

The most handsome piece in this area is undoubtedly the pierglass-console combination by the front door, dating from 1810–1815. The table is somewhat plain, with wooden columns having carved necking and bases. The drawer handles are of bronze, and the rear, flat legs have applied bronze decorations. The pier glass is of an architectural type, with reeded columns and acanthus-carved capitals, the architrave blocks being adorned with oval paterae and, in between them, an acanthus-patterned architrave with a molded cornice topped by a deep cavetto molding—altogether a distinguished-looking piece.

Normally, the console top is occupied by a china bowl of flowers from the residence gardens, competing for attention with the two crystal-hung Italian candlesticks that share the space. These latter incorporate bronze figurines and bronze-bedecked columnar sections in white marble, with black-marble bases.

The whole of this area is illuminated after dark by a large, four-candle (now electrified) lantern of the hanging globe type, actually a reproduction but nevertheless quite effective as an ornament.

On the main floor, the guest bedroom largely follows the pattern of those elsewhere, the major variation being in the color schemes, and the style more or less adhering to the Late Empire period. An excellent example is the enormous mahogany bed which was made around 1860. The posts are huge, being patterned from clustered columns on heavy, square legs, chamfered at the corners, and held together by extremely deep side- and footboards, high enough to call for the use of bed steps when getting either in or out. There is also a very high carved headboard which helps keep the structure vertical— a wise precaution with that massive canopy frame built up of wide and deep moldings from which a lined valance of cotton print depends, matching in pattern the dust ruffle reaching from the foot- and sideboards to the floor.

Outside this bedroom door is an equally massive tall clock, made in Cincinnati in the early 1900's. This stands as high as the doorframe, with its hooded top supported by Ionic pillars at the front. Actually this is a handsome piece of its kind, with a finely filigreed dial, gilded and ornamented, and including a semicircular face indicating moon settings for both North America and Europe. The striking mechanism works on tubular gongs, the largest of which is about 5 feet long, while the weights which operate the mechanism account for 50 pounds of its total avoirdupois.

Reserved to succeeding presidents of the University of

Just inside the entrance is this lovely pier-glass-console combination.

In the rear hall is a mirrored stand, a metal bush, a carved sconce, and an Empire sofa beneath an Italian painting.

Georgia, this house is a splendid survival of a period in American architecture when the classical had reached its peak and the stodgy Victorian, elsewhere, had begun to supersede it, as it did in the field of furnishing. The interior designers, working with the accumulated furniture pieces of several generations, varying in period and provenance—American, French, and English—were able to reproduce to perfection this series of eclectic interiors. They re-created a mid-nineteenth-century atmosphere that did full justice to the lovely exterior while providing all the elements of a thoroughly comfortable home combined with those that made possible its use for important social functions, thoroughly representative of the distinguished institute of learning which it serves.

This beautiful mahogany bedstead, *circa* 1860, dominates the guest bedroom.

OPPOSITE: Twin chandeliers and fireplaces, and an elaborate frieze, grace the drawing room.

STAN HYWET

Renaissance England in the Midwest

One of the most fascinating eras in the history of English architecture was the Tudor period, beginning in the late fifteenth century when the medieval moated manor was giving way to the unsymmetrical jumble of gables and turrets that heralded the free-and-easy style which began to blossom in the reign of England's first Tudor king, Henry VII.

Since rooms of all sizes and shapes could be added at will, and picturesqueness could be achieved while securing the maximum of convenience and comfort inside, it is no wonder that the straggling Tudor should oust, in so many instances, the French château and Italian *palazzo* as favorite of the more genteel and less flamboyant of America's wealthiest families in the early-twentieth century when personal fortunes reached fantastic levels. Happily, this coincided with a period when almost every modern Croesus was also an amateur antiquarian and could decorate his mansion not only with 400-year-old furnishings, but also with treasures from all over the world.

Such a man was Frank A. Seiberling, noted American industrialist, the design of whose Tudor-style brick house of sixty-five rooms (called Stan Hywet from the Saxon words for "hewn stone") is based upon that of several early English mansions with which he and his architect were familiar. The most important of these country houses were Ockwells Manor near London and Compton Wynyates in Warwickshire, the latter a one-time thirteenth-century fortified manor house rebuilt by Edmund Compton between 1481 and 1492, during the reigns of Henry VII and Henry VIII, and continuously owned by that family for the past 700 years.

One important feature of the Seiberling house that was adapted from Compton Wynyates is the entrance porch, which was modified, in the American version, by substituting a half-timbered gable with carved bargeboards for the original crenellated parapet. Next to this porch stands a four-story, battlemented stair tower, the rest of the building being limited to three stories, many of the roofs having doghouse dormers to

The entrance porch with its carved bargeboards alongside the battlemented stair tower.

OPPOSITE: The mansion's largest room—the music room and occasional ballroom—paneled from floor to coved ceiling.

The stunning entrance gate, its carved stone lions displaying shields normally adorned with the family coat of arms.

Stan Hywet (carved stone)— an authentic Tudor reproduction.

light the top floors, together with tall, clustered chimneys that emphasize the multiplicity of roofs and gables of varying height.

An attractive feature of all the elevations is the variation in texture from red, creeper-clad brick, to gray sandstone, to time-darkened timbers and heavy slate, with agreeable patterns formed by the carved doorways and the windows with their small diamond panes, both flat-topped and arched, which display characteristic sixteenth-century eared dripstones and heavy stone mullions. The house is set amid tall, ancient trees and pyramidal evergreens, so that it seems to have been part of the landscape since time immemorial.

Inside, the architectural details are sometimes overwhelming. They are as accurately reproduced from medieval originals as the imported English craftsmen could make them, though some of the paneling and stone fireplaces are background features removed intact or remodeled from centuries-old English castles and manor houses no longer in existence. There is so much of this borrowed detail, around which certain Stan Hywet rooms may have been designed, that descriptions of individual art and craftsmanship must be relegated to a discussion of the particular room in which they are found.

The three-story great hall is an inspiring sight in its modern dress, combining the old with the new, and though exceeded in size by the ballroom, it is the only room without a ceiling. At one end it has a tremendous fireplace, beautifully carved above the Tudor-arched opening, and with its sides terminating in tall Saxon columns. Above these the sloping head tapers back to the chimney breast, which soars upward to wall-plate level, where the ribbed vaulting, the braces and collar beams, the rafters, and even the half-timbering of the gable are visible in the open roof.

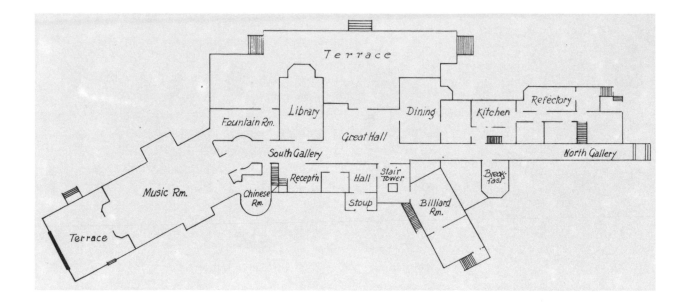

Supporting the main timbers are carved stone corbels on a level with the twin arches of a small window (reminiscent of a medieval "squint") opening into an upper room from which the entire spectacle below can be observed. Wainscoting of sunken panels covers the walls to half their height, with rough plaster above them, as well as in between the rafters. The huge mullioned windows are sunk deep into the thick stone and brick walls, enframed by corbel-high, four-centered arches of wood with a decorative architrave. Inside the window reveals the wainscoting is returned, and the arches' soffits are coffered in wood, one of the windows having a lantern hanging from its center.

Each of the half-story-high windows of leaded glass is divided by stone mullions into nine diamond-paned sash—the upper three arched—six of them incorporating colored heraldic devices of English noble houses. The floor of this room is covered with flagstones set in cement, a treatment accorded elsewhere only to the hallways. Facing the windows is an open mezzanine supported by molded posts, fancifully named the minstrels' balcony, though the early minstrels' "gallery" was normally at one end of the hall, and quite small.

In this hall, as elsewhere in the mansion, the furnishings consist largely of antique pieces and decorative items acquired by Mr. and Mrs. Seiberling during their English and European trips, the majority of them rarities of some importance, plus certain modern additions needed to supply the element of comfort missing in houses everywhere of 300 or 400 years ago. However, so skillfully has this been done that there is not the slightest sense of incongruity in such things as finding a huge Oriental rug on the stone floor, or the fireplace flanked by enormous overstuffed sofas in rose velvet consorting with an Elizabethan melon-bulb table and a Reformation survival—a praying chair shaped like a bishop's throne, with a drop panel forming a Bible rest at kneeling height and exposing a crucifix.

The upper walls here are draped with a pair of sixteenth-

Studied irregularity is a feature of the plan that gives the house its fifteenth-century character.

The magnificent staircase with its strapwork carving and triple-tiered stained-glass windows.

The "praying chair"
in the great hall.

The hall's rare Gothic ceiling
and linenfold paneling.

century Flemish tapestries, while the windows are hung with long draperies of velvet brocade over sheer glass curtains. Of special interest is a beautifully carved oak-press cupboard, dated around 1600, on top of which are perched a pair of bellarmine seventeenth-century stoneware wine jugs (a German import in those days) and an earthen water jug almost as old.

Under the nearby window is a Jacobean gateleg table, displaying a Georgian silver tea set on its huge galleried tray. Nearby are several joint stools used either for seating or as occasional tables, as well as a highbacked French *caqueteuse* and several English turned chairs of the early eighteenth century, which, strangely enough, do not look amiss alongside a polar-bear rug.

A stag's head with widespread antlers, a golden sunburst clock, and a pair of oil paintings adorn the chimney wall—the portraits being of Lord Huntingdon by George Romney (1734–1802) and Sir Thomas Buckler Lethbridge by Sir Thomas Lawrence (1769–1830). More mounted heads are spaced along the gallery rail, interspaced by wrought-iron *torchères*, the latter far outshone by an enormous central chandelier of the corona type once used in churches. These are composed of pyramidal rings of tapers, the whole fixture being lowered by chain or rope for lighting. To the right of the great-hall fireplace, an almost undetectable door in the wainscoting gives access to the library.

This library is a different kind of room from the great hall and, though somewhat smaller, is so exquisitely proportioned that it does not need to be monumental in order to be impressive. Furthermore, it has a beautiful ceiling and walls that create an atmosphere of cozy loveliness whichever way one looks. It is these architectural elements that create the setting in which the furnishings look their best, unifying the whole interior so that it is quite impossible to decide what feature is more, or less, important than the rest. The ceiling, for example, has form, pattern, texture, and color, attributes which separate it from its purely functional aspects. The sturdy beams and the joists they support have their lower edges beveled, molded, and decorated. These timbers are painted to represent walnut, the joists being outlined in gold, the beams beaded in black with a painted pattern along their undersides that looks like low-relief carving. Between them the ceiling is painted in decorative panels of Italian design that give it the appearance of embossed leather. The upper walls are papered in an Oriental pattern of yellows and browns down to the paneled wainscoting. Over the wide arched doorway, however, a recess in the upper wall is bridged by decorative strapwork carved out of wood, its curves and loops ornamented with turned finials, and each end terminated by twin obelisks that crown the tops of the door-arch's flanking twin pilasters.

The double doors—oak on the outside, walnut inside—are set within the recess and are paneled to match the wainscoting, but they have, in addition, a central upper panel

Modern touches fail to spoil the breakfast room's antique air.

carved to represent a decorated arch in the Renaissance manner. These arches are repeated in the elaborately carved overmantel, a larger panel between them enframing a cartouche with a grotesque head in its center. These three panels are separated by high-relief terms with female busts, surmounted by column capitals over which carved modillions support the cornice, exposing between them carved sections of the frieze. The terms rest on carved plinths, separated by decorative panels, under which a massive carved mantel beam forms a narrow shelf on which the only piece of garniture is an octagonal gilt-bronze clock on a balustered and footed base.

In this room the windows are of leaded glass, the central panels reproducing antique bookplates of famous Americans, and in each window recess is a cushioned seat set against the lower wainscot panels. Above the windows the wainscot frieze and cornice are continued around the room, including the area next to the chimney breast where the paneling gives way to bookshelves and low cupboards. Incidentally, one section of the bookcase swings open to give access to the great hall. The flooring is of teak with rosewood strips between the planks, all tied together with butterfly wedges and pegged to the joists. Over this floor is an almost room-length rug in dark green with a wide border in red, green, and tan.

The lighting here is well dispersed, the main sources being two eight-branched globular crystal chandeliers, plus two English-style metal *torchères* converted to floor lamps, located on either side of the entrance archway, and a pair of table lamps with huge porcelain, covered vases and bases, and gold, parasol-

A corner of the master bedroom with its Tudor fireplace and Elizabethan bed.

171

A corner of the three-story great hall with its wainscoting and tall mullioned windows.

like fringed shades. The latter stand on a splendid pair of English half-moon tables, *circa* 1700, which serve the modern back-to-back sofas in gold velvet centered opposite the fireplace. Other important pieces include a Jacobean desk and chest, a pair of English armchairs, and a leather-topped library table with carved legs, such items lending the room a luxurious air that is eminently suited to the rich and splendid pieces, old and new, with which it is furnished.

The largest room in the mansion is the music room, which also serves, on occasion, as a ballroom. It is completely paneled from the floor to the springing line of the coved ceiling, the latter painted a creamy white and intricately compartmented by molded ribs of stucco. A huge fireplace is set well into the wall and is flanked by twin fluted pilasters, the overmantel consisting of the sunken wainscot panels. The fireplace facing is of carved limestone with its arch slightly modified from the standard Tudor pattern.

The entrance from the south gallery is bridged by a wide, flat arch with decorative panels on either side and what might well be a minstrels' gallery, or even an organ loft, above it. Closer to the side walls are smaller doorways with leaded lights, leading from the fountain room and the Chinese room respectively. At the opposite end of the music room is a wide, semioctagonal recess, up two shallow steps so that the raised floor forms a small stage. The rear wall of this and its central door are almost wholly of dark, leaded glass, behind which is an enclosed terrace. Surrounding the proscenium arch is a wide band of French tapestry depicting the four seasons; the stage curtains behind them are of a rich scarlet velvet. On either side of this, the end wall panels have decorative jigsaw cutouts which act as a mask for the organ pipes without cutting down the volume of sound.

The floor is of oak planking, largely covered by a huge

The library with its elaborate carving and painted ceiling has an atmosphere of cozy loveliness.

Persian rug in a tree-of-life design in blue on red. And it is this lively but rich red coloring which dominates the room, not only in the draw curtains but also in the Scalamandre draperies and pelmets, the upholstery of most of the seated pieces—and even the robes of Lord Craig in the overmantel portrait by Raeburn.

There are only three other paintings in the room, portraits of Major Peirson by Romney, Lady Strange by Lawrence, Mrs. Musters as Hebe by Reynolds—a deficiency counteracted by rare musical instruments hung on either side of the fireplace. Three enormous chandeliers of English crystal with pear-shaped amber drops provide the major lighting, assisted by ten crystal sconces and two Italian *torchères* flanking the stage—more than sufficient to counteract the light-absorbent qualities of the dark paneling, thanks to a highly reflective ceiling.

Throughout this room the furniture represents a collection of elegant antique pieces, well exemplified by a beautifully banded harpsichord said to have been played upon both by Handel (1685–1759) and Haydn (1732–1800), and a richly decorated Spanish vargueno of the mid-sixteenth century on its stand. Most of the chairs and sofas, both low- and high-backed, are sixteenth- and seventeenth-century English, including the fireplace bench with its elaborate stretchers. The upholstered pieces are mostly covered in either red velvet or tapestry. Of special interest is a French fourteenth-century prayer bench which was used by six members of the Seiberling family at their weddings. Other striking pieces include a pair of Chinese Coromandel chests and an elaborately carved Italian *cassone*. The organ manual, incidentally, is encased in a paneled walnut cabinet in an antique mood that is unobtrusive against the paneled walls.

Stepping from any of these great rooms into the south gallery is like entering an old abbey where everything is coldly formal though exquisite in execution. Here, between the cold stone floor and the delicate fan vaulting above stands some of the loveliest linenfold paneling in the darkest of oak, topped by narrow strips of intricate carving two panels wide.

In this stately hallway are Venetian lanterns, iron, basket-type hanging lamps shaped like braziers, and twin sconces of iron, all casting a "dim, religious light" on the few pieces of medieval furniture, such as a sturdy, sixteenth-century Italian folding chair in walnut and leather and a rare four-footed wainscot chair, 5 feet long, with a distinctly ecclesiastical air.

There are many other features in the wonderland of Stan Hywet—now the property of Stan Hywet Hall, Incorporated—worthy of detailed description, such as the magnificent staircase with its strapwork banisters and stringer border; the raised fretwork posts; its triple-tiered stained-glass window; the old grandfather clock at the second turn; and a hundred other details that add up to a splendid whole the like of which we shall probably never see again.

The fireplace end of the great hall overlooked by a twin stair window.

San Mateo, California

FILOLI

The House of Flowers

At the rear round-topped French windows are encircled by trained creeper.

Set down in a rolling, well-wooded landscape half a mile from the highway, Filoli is approached along a winding carriage driveway through gradually thickening foliage until, rounding a final bend, one suddenly finds oneself in the wide gravel-surfaced courtyard of this hidden Eden—the whole area ringed with precisely clipped, foot-high box hedges and young flowering magnolias. On each side of this open space is a creeper-clad wing of the two-story mansion whose Doric-columned, stone-balustraded portico peeps out demurely between tall cypresses overshadowed by the red-pantiled roof and massive chimneys of the house itself.

From this vantage point there is little more to be seen, but a sortie on foot quickly reveals the enormous extent of the great lawns and gardens, formal and informal, extending in all directions. Here hedge-lined brick and gravel walks are interspersed with tall, flat-topped Irish yews, spreading stone pines, giant cypresses, and flower-drenched bushes, the beds of blossoms dotted with standard roses and ball-shaped topiary masses of boxwood, with an occasional sundial or fountain at the junctions and, in the distance, the sunstruck roofs of the greenhouses whose contents contribute so much to the perfection of the interiors.

At its rear, the great house reveals a vast area of warm brickwork free of the trained creeper which climbs up on either side of the central, round-topped French doors to cling to the iron balconies above, one whole story beneath the roof cornice with its row of stone-carved modillions, and there again are the round-pedimented dormers and the stone-capped, arched brick chimneys.

This, then, is the U-shaped, forty-room mansion contrived by architect Willis Polk in 1915 and completed three years later. But it was not until 1938, when the estate became the property of Mrs. William P. Roth, an avid horticulturist, that the gardens began to assume a dominant role in the development of the estate comparable to that of the Roth stables whose

OPPOSITE: The walnut-paneled library fireplace and its exquisite seventeenth-century rug.

The entrance to Filoli flanked
by foot-high box hedges
and flowering magnolias.

trophy room is one of the wonders of the horse-show world. Today those gardens and hothouses keep the mansion filled with fresh-cut blooms, day after day, whether the owner is in residence or not.

No more impressive demonstration of flowers as decoration could be imagined or one more compatible with the architectural features of an interior than that afforded by the otherwise unfurnished entrance hall, or foyer, of Filoli. Stepping through the great sidelighted and fanlighted entrance portal into this oval room with its white-marble floor, limestone walls, and high-domed ceiling is an exercise in pleasurable amazement.

This curvilinear space, illuminated by long, iron wall sconces, is decorated entirely with flowers—potted blooms banked against the walls and massed in giant *cachepots* on tall black-iron stands, filling the air with perfume and the room with gorgeous color which looks its best against the pale stone. Opposite the main portal are huge double doors of mahogany giving access to the inner hall, the arch above them glazed to admit light and keep out sound and cold air, while to the right, on the long axis of the room, is a rectangular recess accommodating an arched opening sealed by twin, quadrant-topped doors. These lead to a powder room where a circular central table, Roman in style, with caryatids for legs, together with several small side tables, support silver and porcelain containers of flowers—the whole a spirited and quite unusual introduction to the treasure-filled rooms beyond.

The inner door from the foyer opens into the great transverse hall which runs the entire length of the mansion from the kitchen wing to well beyond the great stair hall to a garden door masked by voluminous crimson draperies. Three pilastered arches divide the transverse hall into four sections, the first and third of which, looking toward the service wing, have barrel ceilings, one of the two with flat ceilings opening into the wide stair hall, the other into the drawing room opposite the foyer door.

Facing the stair hall is a magnificent doorway, typical of many in this great house, which opens into a private sitting

Filoli's domed entrance hall.

Filoli

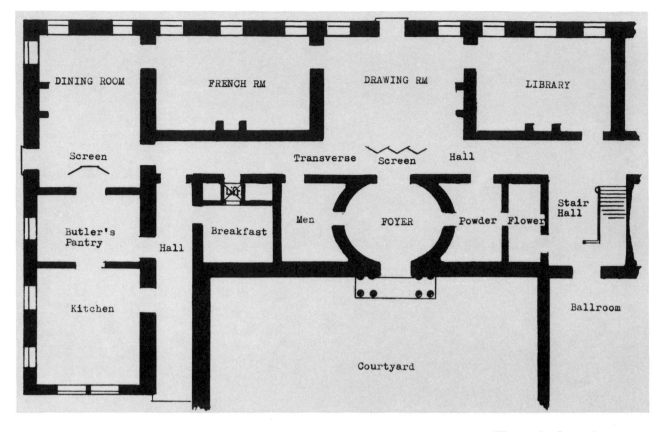

The main-floor plan,
showing bedroom elevator.

The stair hall leading to
the ballroom doorway.

room. This doorway has both a Corinthian pilaster and an attached column of the same order on either side of the opening, under a pillow-block-style entablature and a swan-neck pediment with beautifully carved scrolls. These extend upward to within a short distance of the handsome frieze and cornice, which are carried around all four sides of the ceiling area that is open to the stair hall.

Flanking this doorway is a pair of half-round pedestal tables in mahogany, each with its large blue Ming vase. In the stair hall itself, the wide parquet floor is largely hidden by an antique Karabagh floral carpet in a huge palmetto pattern, somewhat similar to that of another Karabagh in the adjoining transverse hall.

The stair hall is big enough to constitute a room in itself and is more than attractive with its "easy" staircase with steps of black marble carpeted in blue, its filigree-style, black-iron balusters and shining rail sweeping down into a wide loop matching that of the curving second-step end. Over the stair hangs a large English Mortlake tapestry illustrating a falconry group on horseback and other bucolic seventeenth-century activities, while in the back wall of the stair hall a wide, arched doorway, with a double door, leads to the ballroom, which forms one of the wings. A small doorway nearby connects the hall with the Flower Room, behind which is a powder room accessible from both the foyer and the transverse hall.

In the long transverse hall the several arches and ceiling variations not only break up the perspective, but also facilitate the introduction of interesting furniture arrangements. Looking in one direction from the stair hall, for instance, a narrow Oriental runner allows bare floor space at each side for a beautiful eighteenth-century Queen Anne burl-walnut drop-leaf table, with its bouquet of carnations and baby's breath, to face a brass-handled Chinese traveling chest in an area where the painted walls are enlivened by antique needlework panels. In the other direction is a carved Chippendale side table adorned with an antique birdcage, while, at the point where the drawing room opens into—and absorbs—the hall, the wall is covered in an old gray-green Italian damask, with the foyer door in the middle. In another section is an eighteenth-century Italian table with an inlaid marble top, carved knees, and *pieds de biche*, flanked by a pair of Queen Anne cabriole-legged chairs.

Spaced from the foyer door by the width of the transverse hall is an exquisite, tall, six-leaf, painted French screen depicting rural scenes. This serves not only to divert possible drafts from the opened door but forms a delightful background for a furniture grouping in the drawing room proper, the most formal room in the house.

In this room the gray-green of the wall covering is matched by the frieze and cornice, a color emphasized by the off-white ceiling and the cream trim. These walls tie in beautifully with the gray and green draperies on a beige background picked

The transverse hall at centerpoint.

An eighteenth-century Dutch *bombé* cabinet in figured walnut.

179

The ballroom fireplace of ormolu-decorated marble.

The ballroom itself forms a separate wing.

out with gold thread, the pelmets being of the same material edged in green, all over glass curtains of a pale-green silk. In contrast, the parquet floor, in the Louis XV classical mosaic style, is covered by a huge Feraghan carpet in a pattern of light blue, reds, tan, and black. Architectural touches are supplied by doorways in two facing corners of the room, duplicates of the scroll-topped hall design described earlier, together with the carved modillions of the cornices.

At one end of the room is a white-marble fireplace whose attached Ionic columns and Greek-key frieze, centering on a carved panel, are picked out in gold. The mantel garniture includes a pair of bronze-mounted celadon vases, over which is hung a notable sixteenth-century *mille fleur* tapestry with a central heraldic device, probably woven in England.

Against the opposite wall is arranged a long library table flanked by a pair of carved and gilded *torchères* surmounted by eight-light crystal candelabra with amethyst drops. At the center of the table, a fine eighteenth-century bracket clock by William Webster is mounted on an elaborately carved, ivory-inlaid Chinese stand between two marble figurines—one of Apollo, the other of his son, Orpheus, god of music—a grouping completed by a pair of large, green, Ming temple jars in a simple bamboo pattern.

Tying together this exquisite display is a wall-hung sixteenth-century French tapestry of silk and wool in greens and tans—the whole constituting a glorious backdrop for a sofa and two Chippendale chairs in crimson damask, the latter astride a Chinese coffee table in black lacquer outlined in gold whose

top finds room for a live bouquet of carnations and baby's breath, while two other, similar pieces with filigree-decorated panels serve as end tables to the sofa—the whole a stunning arrangement that no fireplace could excel for charm and interest.

Intermingled throughout this room are some of the loveliest furniture pieces ever devised, the varying colors and fabrics acting as foils, one for the other. For example, in front of the wonderful painted screen in the colors of outdoor nature are grouped a Jacobean-style gateleg table in ebony, flanked by a pair of tall-back chairs. One is a magnificent Cromwellian-style wing chair in red and gold. The other is a Louis XV flat-back version, with cabriole legs and scroll feet, in an equally fantastic but far less regal floral needlepoint in green and silver with a touch of pink on a muddy brown. The ensemble centers on a huge silver goblet full of golden, buttonlike chrysanthemums. Elsewhere the contrasts are simpler and subtler, with large areas of crimson damask consorting with pale-pink flowers woven into a paler blue fabric, and Chippendale chairs in red with heavy, tasseled fringes.

Another interesting feature of the room is the arrangement of the furniture into conversation groups with a multiplicity of occasional sofa and coffee tables—each with its quota of flowers, all readily rearrangeable to accommodate larger, homogeneous groups.

Adjoining the drawing room is the warm and inviting library, which is paneled in walnut and has a white-marble fireplace. The upper paneling has a decorative molding and is shaped to surround the open bookshelves, the plain panels below forming a dado which extends around the room.

The floor is almost wholly covered by a huge seventeenth-century rug (a Polonaise museum piece) with a crimson field whose design in pink, cream, and light green forms a stunning background for one of the finest furniture pieces in the house— an authentic Chippendale pedestal writing desk once in the possession of the Countess of Warwick. Richly carved in the Gothic manner, it also has panels of fret ornament. The top is inset with dark-green leather, the edges carved, and the drawers equipped with elaborately chased drop handles in gilded bronze.

Besides the colorful bindings of the books, other colors and tone values are introduced by the blue-green of the sofas, the red and blue coverings of the Chippendale, Sheraton, and other English chairs, all skillfully distributed in order not to bunch the hues, and the black lacquer of the assorted small tables against a lively pale-blue background afforded by the silk-damask draperies. These and the many antique pieces in mellow walnut and mahogany tones contribute a note of quiet elegance to the room, whose lived-in look is emphasized by the lovely bouquets of flowers on almost every table.

On the other side of the drawing room is the French Room, so-called because of its fine collection of French mezzotints,

The dining room is illuminated solely by candles.

A drawing room setting around a sixteenth-century French tapestry.

A feature of the French Room is this exquisite boulle chest.

Another French Room piece— a marble-topped boulle chest in *chinoiserie*.

The library's magnificent Chippendale desk.

paintings by French impressionists, and predominantly Louis XV and XVI furnishings, although in recent years it has also acquired a variety of English examples in both categories.

This is quite a large room, with a fireplace on the long wall facing the three windows, its apparent proportions being somewhat influenced by the deep, cream-colored frieze and dentiled cornice and the division of the beige-colored walls into tall, narrow panels by the application of two-inch-wide gimp in gold and white. The floor here is also of parquet, following the drawing room's Louis XV pattern. Because of these backgrounds and the height of the ceiling, major eye-catchers are the room's twin crystal chandeliers of a light and airy Louis XV design with bronze *bobêches*. On the walls are equally delicate crystal sconces.

The basic colors in this room are provided by the draperies with their rather formal design of flowers in pastel tones on a background of gold. These are tied back by tasseled cords and finished off with deep, tailored pelmets bordered with multicolored, tasseled braid. The principal rug, which is worthy of display on a wall, is a fine, late-sixteenth-century Isfahan lined with taffeta. This has a mulberry field with a border of dark blue. Two other Orientals of minor interest provide darker accents, one in front of the fireplace, the other fronting the door to the dining room. The fireplace is small and neat, with a mantel of carved white marble around a facing of black slate and console-style pillars. Over it is a perfectly scaled Adam mirror in blue and gold which matches the delicacy of the small, gold-framed prints that line the walls, while on the mantel itself parades a round dozen of Chinese glazed figures depicting "Moods of the People"!

Among the major pieces of furniture are the twin Louis XVI marble-topped console tables in white and gold along the window wall, with their companion Rococo-style mirrors that are tremendously effective with their deep, carved, and gilded crests and aprons. Contrast also is supplied by an odd pair of white and gilt oval-backed *fauteuils* in petit point and Directoire armchairs in pale blue, with painted fruitwood panels. However, far more interesting—and valuable—are three other important pieces: a marble-topped chest extravagantly decorated in *chinoiserie,* with ormolu mounts; an eighteenth-century French marquetry commode with a Greek-key frieze, bronze decoration on the quarter posts, and ormolu sabots (incidentally signed, "L. Peridiez JME 1761"); and the comparatively modest commode by Chippendale, banded in harewood, its lacquer panels enlivened with painted flowerets.

Between the French Room and the kitchen wing is the dining room, which has the distinction of being illuminated after dark by candles, the only truly atmospheric—and romantic—source of light. In this instance the chandelier is a Louis XIV pattern in bronze with twenty-two tapers, augmented by matching wall sconces with five candles apiece.

In this room the already high ceiling is coved and painted

A corner of the French Room emphasizing
the parquet floor.

off-white for added reflection as demanded by the walnut-paneled walls, which are relieved, insofar as light absorption is concerned, by a huge fireplace and hearth of streaky white marble and, on the adjoining kitchen side, by an even larger three-panel screen over the kitchen doorway. This is an English piece of the George II period, 8 feet by 6 feet, covered in a Fulham tapestry of silk and wool over a carved base.

Between the two end windows of the room stands a *bombé*-style Dutch cabinet in figured walnut, its upper part forming a delightfully shaped vitrine, in which some of the better glass and chinaware are displayed, including crystal goblets and Spode and Worcester tea and dinner pieces. In front of this stands the large dining table, the smaller, family one, which is a sixteenth-century circular Italian piece in walnut, being temporarily located under the chandelier. Around it are normally grouped four of the room's dozen handsome William and Mary-style walnut, high-backed, tapestried chairs, depicting each month of the year, which are a colorful addition to the room. Other details include a seventeenth-century William and Mary embroidered fire screen and an overmantel oil painting of hanging game by Dutchman Jan Weenix.

Such are the Filoli interiors—a name coined from the motto "Fight bravely, love bravely, live bravely"—a house that has a unique personality thanks to its owner's capacity for blending so successfully the skills of the world's finest artists and artisans with nature's most exquisite gift—the "flowers of the field."

A game table in the library.

Annapolis, Maryland

THE HAMMOND-HARWOOD HOUSE

William Buckland's Masterpiece

The Hammond-Harwood
House—garden front,
showing window-door.

In 1773–1774 William Buckland, the English-born architect of Gunston Hall and other important residences, designed and built this mansion in Annapolis for lawyer-plantation owner Matthias Hammond—a town house that turned out to be a masterpiece of Late Georgian design and, since he died soon after its completion, Buckland's *chef-d'œuvre*. Today the mansion is known as the Hammond-Harwood House, the names of the first and final families to own it.

The amazing thing about this house is the number of structural and decorative details that have gone into securing the finely balanced exteriors and attractive interiors, whose elaborate enrichment does not in the least reduce their feeling of spaciousness—or of charm. So far as the design of the building is concerned, the arrangement of the five parts is in normal sequence. This elongated structure, built of brick laid in Flemish bond, relieved by a rubbed-brick belt course and lintels, is composed of the central block—the two-story mansion proper—with end wings, one accommodating the owner's law office, the other a kitchen and service unit. These are connected to the main body of the house by single-story hallways which have exterior doors, front and rear, the office unit being entirely closed off from the house itself. Each of the end wings has a polygonal bay at the front—a rare feature before the Revolution—and the kitchen wing, in addition, has a shallow lean-to along the outer side which forms a narrow corridor connecting the cooking area with the rear service room and incorporates an entrance door at each end.

Although both the main wing structures are two stories high, their roofs are much lower than that of the house, which not only has higher ceilings but also is perched on a raised basement. Among the features that make this façade particularly impressive are the deep flanking bays with three windows on each floor; the pedimented pavilions of the hyphen doorways with their semicircular fanlights flanked by arched recesses in the brickwork walls; the great central pavilion of the house with its modillioned cornice and bull's-eye-adorned pediment fronting the low, hipped roof; the principal entrance with its own pediment, laurel-carved, pulvinated frieze, attached columns, and double door recessed in an arched surround with a semicircular keystoned fanlight, festooned with roses in high relief and other classic details. Above this doorway, the central second-story window is enriched with decorative stiles and brackets and a flat pediment. From sidewalk level the main entrance is reached by six broad, iron-railed stone steps.

Even more unusual is the garden façade on which a pavilion is replaced by a set of four giant brick pilasters reaching from the water table to the white-painted pulvinated frieze. Here the most extraordinary feature is the rear door, which is a duplicate of the rest of the main-floor sash windows, set into a wide wood frame, including a wooden step and a plain overdoor with a molded and bracketed cornice. Below this are five semicircular stone steps, each larger than the one above it, leading down to the lawn. Inside the room this doorway looks just like the rest of the windows, recessed into the thick walls and flanked by folding shutters decorated with octagonal, rosette-carved panels alternating with elongated plain ones. Around this opening is an intricate molding, the upper part of which enframes a blue and white damask valance, forming half-length cascades at the sides, in place of full-length draperies.

Detail of the street entrance.

The mansion with its unusual bay-fronted wings.

185

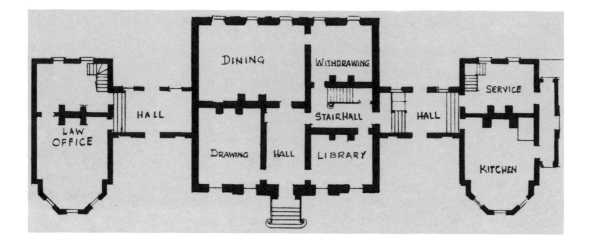

The plan shows the five parts and their uses.

The stair hall off the front hall.

While the rest of the windows continue the dado across the bottom of each recess, and the outer brickwork comes up to sill height, in this instance the dado within the recess forms a pair of doors below the sash. By raising the sash and opening the dado doors outward, egress is possible—a more important feature than it may seem, as will be obvious when the floor plan is examined. This window-door, being in the center of the rear façade, is exactly opposite the front door which opens into the reception hall. To pass right through the house, therefore, it is only necessary to walk from the front hall straight through the dining room to the exit.

With the dining room being extended behind the reception hall, it is not surprising that it is the largest interior on the main floor. It is also one of the most beautiful rooms, not only because of its nice proportions but also as a result of the decorative details, such as the dado molding and the window trim, the latter having a carved panel superimposed, duplicating the overdoors, together with end brackets supporting a molded cornice a foot or so below ceiling level. These carry the eye to the ceiling with its own deeply projecting modillioned cornice. Other decorative touches are supplied by the fireplace, thanks to its crossetted mantel and bracketed shelf beneath a colorful marine oil painting by Francis Swaine of England (1740–1787), which is set into the stucco-framed overmantel panel.

Both this panel and the fireplace frieze are ornamented in stucco in a manner compatible with the rest of the woodwork, so that in spite of all this rich detail, so very carefully scaled to the room, its appearance of overall simplicity is preserved. The setting is completed by a lovely eighteenth-century Waterford chandelier, a blue and gray Isfahan rug, and some splendid paintings on walls of a wonderfully soft celadon green—a wall color used throughout the house. The oil paintings include a copy of a remarkable 1774 portrait of William Buckland by Charles Willson Peale, showing the architect working on the plans for this very house. Other portraits by C. W. Peale portray members of the Buckland, Callahan, and Harwood families, all of whom had some connection with this house.

The front hall.

There is also one painting by Peale's son, Rembrandt (1778 – 1860), which is a copy of Thomas Sully's portrait of Eliza Ridgeley of Hampton. Naturally, all these add to the room's air of quiet distinction, which is due in no small degree to the quality of the period furnishings.

Quite a few of the furniture pieces in this room were original to the house, including a three-section dining table and a sideboard, both of Hepplewhite design, half a dozen Chippendale straight-legged side chairs with seats in blue and white damask, and two with cabriole legs. There is also a Hepplewhite hunt board of Maryland provenance, as are the pair of half-round tables, while the Chippendale cellarette is from Virginia.

On the hunt board are displayed a pair of Chinese Export fruit coolers and a Fitzhugh platter, and on the side tables, twin covered vases, all in blue and white, while the mantel garniture consists of five pieces of Chinese Export porcelain. Most of the elegant silver is from England.

Off the dining room is the much smaller withdrawing room which, together with the library, is located to the right of the reception hall. Both of these rooms are further reduced in size because of the separate stair hall (at right angles to the reception hall) located between them. The withdrawing room is also comparatively plain insofar as the architectural details are concerned. These consist of an egg-and-dart cornice with a beaded border, a white-painted flat-molded dado, and a wooden fireplace mantel with a dentiled shelf, carved pillow-

Portrait of William Buckland working on the house plans.

Detail of the dining room door surround.

The master bedroom.

block frieze with a central panel, a molded surround, and an eared stone facing painted black. Nevertheless it is endowed with considerable charm, thanks to a gay Kashmiri rug in red and blue on a light-beige ground, which goes so well with the celadon-green walls, accented by mulberry damask draperies, copied from an eighteenth-century fabric by Scalamandre. This same mulberry color is used for an English wing chair of unusual design, the wings extending outside the rolled arms, and for an American mahogany armchair with three carved splats (*circa* 1790) and its balloon-splatted mate. There is also a Sheraton-style side chair with a Gothic-arched splat, hailing from Salem, Massachusetts.

The mantel garniture consists of a Washington clock of the type brought to this country by Lafayette in 1824 and a pair of lidded Chinese Export vases. The side wall, however, is almost wholly occupied by a Maryland inlaid, mahogany, breakfront secretary-cabinet, *circa* 1780, which has Gothic glazing bars. Next to it, in the window corner, a delicate spade-footed tripod table displays a lovely tulip-pattern Chinese Export bowl, while facing the fireplace is a D-shaped Hepplewhite-style side table of Baltimore provenance, beneath an 1804 gold-framed watercolor of *Kitchen Locks,* Harper's Ferry, signed "Luke Robbins."

The only other painting is a portrait of William Pinkney (1764–1822) attributed to the French artist Bouche, who had a painting school in Baltimore; but there is some excellent silverware displayed, including a teapot by John Langford of London (1763), a George III inkstand by Sebille, and Sheffield candlesticks of exquisite design whose maker is unknown.

Both the reception hall and the stair hall are of unusual interest, largely because of their effect on the general circulation through the interior and also their separation of function. Access to the stair hall is effected through a doorway to the hyphen connecting the house service wing, as well as to the reception hall and to the library.

The house's main entrance is a double, raised-paneled door in a round-topped frame complete with fanlight and a huge brass lock, all under a 12-foot ceiling and its heavily dentiled cornice, the walls painted the same celadon color as the rest of the interiors. Close by are doorways opening, respectively, into the parlor and the library. All the doors have six fielded panels and simple molded frames with corniced overdoors giving them height and importance, to counteract the effect of the chair-rail moldings and deep baseboards.

To one side of the entrance door hangs an aneroid barometer; on the other a birdcage-type English clock dated 1650–1670, with a long, exposed pendulum and weights, while from the ceiling hangs a George III-type glass bowl lantern. Down the center of the pine plank floor is a long Feraghan rug, flanked by a mahogany Maryland hunt board, *circa* 1810, attributed to John Shaw, and a Maryland marble-topped side table of about 1790, with an inlaid apron.

A corner of the library.

On the hunt board is a fascinating punch bowl of Chinese Export porcelain decorated with hunting scenes and a pair of 1755 columnar silver candlesticks by William Cafe of London. Over this hangs a large portrait of Philip Yorke (1690–1764), First Earl of Hardwicke and Lord Chancellor, in its eared frame, the artist, alas, unknown. The marble-topped table is adorned with a pair of hurricane globes dated 1800, shielding Sheffield candlesticks beneath a splendid Chippendale looking glass (*circa* 1790) in parcel-gilt mahogany. Seating is confined to a pair of Philadelphia Queen Anne walnut chairs with slipper feet and a cane-seated and painted oval Maryland bench.

In the smaller stair hall, alongside the Kabistan rug, is an English Hepplewhite bench under a looking glass of the same provenance, dated 1740. The stairway itself is also quite simple, the sweep of the square-balustered handrail repeated in the wall molding above which hang some of the ten engravings and mezzotints of English eighteenth-century notabilities—and one of Trumbull's *Bunker Hill* decorating the hall.

Occupying a corner of the first stair landing is a Hepplewhite-style tall clock in a Baltimore-type case, made by George Long of Hanover, Pennsylvania. Just above this is the second-floor round-topped window, within whose paneled reveal is displayed a 1740 Chinese Export urn with pistol-grip handles and a painted medallion. Overhead, a deeply coved ceiling soars upward into the roof. On the second floor is one of the loveliest—and largest—rooms in the mansion: the ballroom which is directly over the dining room and doubles as a music room, as its normal style of furnishing will suggest. Strangely enough, the decorative plasterwork is confined to the frieze and cornice, the design consisting of groups of six beaded rods alternate with Adam-style urns, having a tiny floral band below and dentils above, all under a neatly detailed cornice. Though there is a splendid Waterford chandelier, the ceiling is quite plain, as are the door surrounds. On the other

Some of the exquisite game room furniture.

The sparsely furnished
withdrawing room.

The second-floor ballroom.

hand, the fireplace on a projecting chimney breast is beautifully decorated with applied floral and ribbon moldings, and the eared facing is encased in a carved and beaded molding.

Over the mantelshelf is a portrait of Mrs. Ninian Pinkney (nee Hambleton) of Easton, done by John Peale Bordley (1800–1882), accompanied by a pair of crystal candelabra. The chair rail is of heavy but simple molding carried into the deep window recesses, which are quite plain except for the inside shutter enclosures, in contrast with the highly detailed and painted pelmets from which hang the extraordinarily rich tasseled swags and draperies of eighteenth-century French design in gold and ivory lampas. Between three of the windows hangs a pair of Georgian carved and gilded mirrors (*circa* 1735), which have scroll pediments and Prince of Wales feathers for a crest.

On the room's pale-gray Aubusson rug is an English Broadwood grand piano, *circa* 1806, with an Everard harp and its 1840 music stand nearby, together with a stool whose seat is covered in blue and ivory silk brocade. The remaining seated pieces include one Hepplewhite and two Maryland Chippendale side chairs with seats in silver and green satin and a Boston Hepplewhite sofa (*circa* 1780) in gold and ivory brocade. These complete an interior which, like the rest of the rooms, though highly individualized, is thoroughly representative of the house's day and age and the quality of the gentlefolk who called it home.

OATLANDS

A Nostalgic Federal with Adam Touches

Reputedly one of the South's finest Georgian houses, Oatlands, which dates from 1803, is actually Federal both in style and period, with a number of gracious Adam touches, its otherwise plain exterior transformed into a thing of beauty by the addition of an exquisite pillared portico in 1827.

In designing his mansion, George Carter, great-grandson of the famed Robert "King" Carter, seems to have adapted many of his ideas from a 1768 book on architectural principles by William Chambers of London. Evidence of this is provided by his letters to the man who carved the architectural details for Oatlands and his sketches for the base and capital of the portico columns, which were made in New York by Henry Farnham, transported to Alexandria by packet, and thence by wagon to the house.

Today these Corinthian columns are the house's principal distinguishing feature. Of incredibly delicate proportions, their shafts are finely fluted, stop-fluted at the base, and topped by capitals that are masterpieces of the carver's art. These stand on a tall, stone terrace above the high basement which emphasizes their twin-story height, and the inter-columnar spaces today are filled with painted flower boxes, filled most of the time—in that balmy climate—with salmon-pink geraniums. In contrast with these masses of color and the green of the boxes and blinds, the stucco walls are a pale yellow; the door and window trim and the columns themselves are white, so that the whole façade gleams proudly in the summer sun.

The main body of the house is rectangular, with two short, semioctagonal wings slightly set back, all two stories high except for the central section the width of the portico which rears up another story. The roofs are flat, and all except the small wings are given added height by solid parapets whose sole decoration is a series of oblong recesses in the vertical face.

Other decorative features of the main façade consist of the entrance doorway with its semicircular leaded fanlight and tall, rectangular sidelights—their pattern revealing an Adam-

Oatlands' famed pillared portico and the mounting block.

The sun-bathed façade with
its five-sided end stair halls.

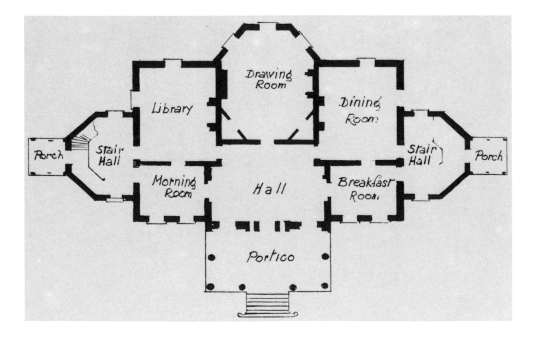

The plan, emphasizing
an octagonal drawing room.

OPPOSITE: The dining room showing china used by
George Washington.

esque touch—all set between four attached Corinthian columns supporting a molded entablature with a dentiled cornice, the center section projecting slightly, *en ressaut*. One peculiarity of these attached columns—or colonnettes—is that the fluting begins at doorknob height, the lower part remaining plain to match the bottom of the door which is only paneled above the lock on the outer face, though the interior side has an extra pair of panels in the lower section. Over this imposing entrance is a Palladian window, both openings being flanked by sash windows with heavy, keystoned lintels. A notable feature here is that the upper windows rise well above the tops of the columns, thanks to the dished portico ceiling resulting from the inside faces of the entablatures being exposed— further evidence of extremely careful design and attention to precedent.

Approached by seven iron-railed stone steps, rising between twin basement windows from a diagonally bricked area, and centered on a marble mounting block, this whole unit is tremendously impressive, setting the stage for the splendid architectural features of the major interiors which have rarely been paralleled elsewhere.

Inside the entrance hall of Oatlands, which is as large as many a ballroom, one is introduced to the interior architectural features in no uncertain manner. The inside face of the door with its Adamesque surround—the stop-fluted pilasters, the delicately molded door panels, the grouped chisel cuts and tiny dentils of the architrave, the triple overdoor panels intricately modeled in low relief—constitutes a totally integrated design in which even the metal swags and husks of the fanlight and the curving leaded lights of the narrow side windows in their paneled reveals contribute their modicum of flowing pattern. Above them all the wall's elaborate frieze, with its egg-and-dart molding, its tiny acanthus-decorated modillions supporting a cornice enriched with a beaded ovolo, stretches right around the jogged ceiling.

Equally imposing, but quite different in design, is the deeply recessed doorway to the breakfast room. Here the pilastered door surround with its blind-arch overdoor is set well within a larger arched opening having a paneled soffit and abutments. Both arches are faced with a tracery band between fillets. In addition, the blind arch over the door opening reproduces, in plaster, the radiating muntins and metal swags of the hall entrance fanlight, though the inner semicircle is replaced by a fan of exquisitely modeled feathers.

From the center of the hall ceiling hangs an Italian brass lantern with crystal drops and rosettes. This is suspended from a huge ceiling rose of molded plaster which makes elaborate use of the Greek anthemion motif, with rosebuds and opened acanthus leaves in the Adam manner—a splendid work of art.

All of these plaster decorations stand out strongly against a pale-gray ceiling and walls of unpainted, sand-finished plaster.

Inside the entrance doorway with its Adamesque touches.

View through the dining room to a stair hall.

195

OPPOSITE: Oatlands' drawing room fireplace today.

The entrance hall with its life-sized portraits of George Eustis and his wife.

The hall's Italian lantern and its Adamesque ceiling rose.

Dominating this room are two life-sized portraits, painted in 1860, one of George Eustis, father of the last owner of Oatlands, the other of his wife, Louise.

In the hall itself one of the most interesting furniture pieces is the Louis XV *bureau plat* occupying the center of the floor. Its tooled-leather top, delicate banding, and ormolu mounts are a delight to the eye. Next in order of attraction are the several Hepplewhite shield-back side chairs (of which there are fourteen in the mansion) displaying the favorite Adam motif, a Greek kylix—a two-handled, stemmed drinking cup—in an oval with festoons decorating the splat. Set beneath the portraits are a pair of mahogany card tables used as consoles to support two lovely Chinese Export bowls.

In the dining room, despite the elaborate entrance hall, the architecture is more restrained yet decidedly pleasing, emphasizing comfort rather than excitement. Here the heavy step-molded cornice spreads itself out over the ceiling, relieved by square rosettes in the angles in the Greek Revival manner. The fireplace mantel also carries out the Greek Revival theme with its attached stop-fluted columns, anthemion-decorated frieze, gouge and punch work, and central panel with its flower-basket carving. Both the heavy chair rail and the baseboard are stopped against the columns. The baseboard, incidentally, is painted black, and the fireplace facing is of black marble, the two contrasting sharply with the cream walls which, in turn, constitute an effective backdrop for the mahogany furniture. The draperies are of a deep Burgundy. Covering the floor is a Persian rug with a gray field and thickly studded with small floral designs in shades of blue and red.

The mantel garniture consists of a pair of decorative urns and a nineteenth-century French clock in ormolu-mounted white marble. Above these hangs an 1825 portrait of Commander Charles Morris, USN, the grandfather of Mrs. George Eustis. The rest of the pictures on these walls are nineteenth-century French hunting prints. The most striking—though hardly the most entrancing—piece of furniture in this room is the massive sideboard, which was made for Mr. George Eustis sometime between 1830 and 1840, together with its large reeded knife boxes. Normally, there stands between the latter an early-nineteenth-century, oval-shaped Chinese Export porcelain, covered tureen and platter belonging to the Carter family, under an early portrait of Governor William Eustis of Massachusetts (1735–1825). Close by, on one of the eighteenth-century side tables, is a huge German punch bowl of pewter made in the eighteen hundreds.

The dining table is an eighteenth-century English pedestal type, with a nicely banded top having a reeded edge, which

"La Vierge d'Autun"—a copy in stone graces the garden.

The sundial terrace and teahouse.

The library with its portrait of
William Wilson Corcoran.

The exquisite terminal piece
in box parterre.

was bought especially for Oatlands. The chairs, however, are part of the same set as those in the hall, leather-covered, brass-nailed, and lovely. The dessert service displayed on this table is eighteenth-century Salopian ware made in the Caughley pottery in Shropshire. Tradition has it that this was part of a dining service used by George Washington at Mount Vernon.

To the rear of the entrance hall is the octagonal drawing room, whose intriguing frieze and cornice again reflect the sophisticated taste of the Adam period. Their whiteness is emphasized by the blue-gray walls, accented by the white chair rail and base molding over a black baseboard. The pillared fireplace is of Pennsylvania marble deeply veined in blue, that intense color being picked up by a porcelain vase used as a lamp base on a nearby round mahogany table. The carpet is a modern broadloom in gray, with antique Orientals as scatter rugs. Other key colors are represented by the draperies of gold damask, a three-paneled folding screen with a floral design in brown on beige, a gilded Louis XV canapé in a dark-green silk brocade, and a pair of *fauteuils* in silk damask of a pale green.

Also prominent in this setting, emphasizing its eclectic character, are a Louis XVI round-back *fauteuil* covered in petit point and a little English, kidney-shaped, three-drawer table (more politely, a *table à rognon!*) with a pierced brass gallery—altogether an inviting, unpretentious room.

The library, next to the drawing room, has the same white, geometrical cornice as the dining room, the walls being a gray-green and the fireplace mantel a contrasting black-veined white marble, while the floor sports a Turkish rug in blues and greens on a soft red ground. Over the mantel, in an elaborate gilt frame, is a portrait of William Wilson Corcoran, grandfather of Mr. Eustis, who gave the Corcoran Gallery of Art to the City of Washington.

Designed solely for private relaxation, this room makes a gesture toward informality in its heavily upholstered sofa and armchair, disguised by slipcovers in a green and gray floral pattern on a ground of off-white. There is, however, included in its furnishings a very nice Sheraton-style round tripod table of 1790, with a wide banding of tulipwood, together with an inlaid card table labeled by Joseph Short of Newburyport, a Hepplewhite lady's writing desk, and a Queen Anne tilt-top table, all in mahogany.

One main-floor room that forms something of a contrast to the foregoing is the breakfast room, which gets most of its character from its colors. This has yellow walls emphasized by the white lines of the trim and the dark red of the draperies, which compete in richness of color with the large Feraghan rug in the Herati pattern. This rug has a vivid design in red, light blue, ivory, and brown on a field of deep blue—a mixture that contributes an air of luxury in which the American Empire mahogany furniture seems quite at home. These pieces include

The west end stairway with its 1882 engravings.

A glimpse of a dining room table setting.

The master bedroom occupied
by Mrs. Eustis.

The Green Room's fireplace
mantel.

an 1835 pedestal table and several quite plain roll-back chairs, *circa* 1820–1830, plus a Chippendale-style wing chair upholstered in a medium green; the whole room is a highly personal setting for the horse prints on the walls and the handsome silver-gilt Meath Hunt Cup standing proudly on the sideboard beneath a crayon drawing of Mr. Eustis.

Still another room with a pleasing character all its own is the master bedroom, which Mrs. Eustis, in the later years, occupied exclusively and therefore furnished in the French style.

This bedroom ceiling is white, as is the beautifully detailed wooden mantelpiece, the door, and all trim, against walls tinted a pale gray. The carpeting is of a delicate blue, overlaid with Oriental rugs in strategic positions.

When the Eustises bought Oatlands in 1903, the gardens were neglected and overgrown, and that section of the former slave quarters in front of the house was in ruins. One of Mrs. Eustis' first projects, therefore, was to remove the ruins which were past restoration and extend the garden in the form of parterres beyond a new balustrade which marked the limits of the front lawn. At the same time, the rest of the outdoor living quarters together with the overseer's cottage were put in good condition.

A few yards away is a well, topped by a marble wellhead from North Africa, safely sealed by an iron grating for the protection of the Eustis children of the original and succeeding generations—a safeguard for visitors, young and old, that doubtless will be retained by The National Trust, new owners of the Oatlands estate, along with the rest of its nostalgic reminders of earlier days and other ways.

Hyde Park, New York

THE VANDERBILT MANSION

Italian Renaissance on the Hudson

One of America's finest examples of Italian Renaissance architecture is the fifty-room mansion designed for Frederick W. Vanderbilt by the New York architects McKim, Mead, and White and erected during a two-year period, 1896–1898, on his Hudson River estate, whose parklike grounds afford unsurpassed views of the river and the Shawangunk Mountains beyond.

Faced with tooled limestone of incredible smoothness, the house is stately without being austere, thanks not only to its delicacy of detail but also to its nicely scaled proportions. Standing three stories high over a slightly raised basement, the tall central block is dominated on each side by a handsome two-story portico whose fluted Corinthian columns and pilasters support a full entablature, giving the mansion an equally striking and impressive appearance from whatever angle it is observed. Ornamentation here includes high-relief lion masks and occasional decorative panels in the architraves and on the upper outside corners of the pavilions which form part of the exterior walls. These straight, smooth surfaces are lovely in themselves, emphasized by the crisp carving of the window moldings and topped by handsome turned balustrades, above which tower the huge rectangular stone chimneys of the house.

On the western façade, overlooking the river, the triple-arched entrance to the rear pavilion is overshadowed by a semicircular portico of six columns, with a flight of iron-railed stone steps leading down from each side to the lawn. The porticoes at the north and south ends of the house extend right across the main body of the structure, each having six columns centering on the steps. The only difference between the two is that the foundation wall of the north portico is pierced by basement windows, whereas that of the south portico is not, a feature that adds to the latter's attractive appearance. In each portico except that of the main (east) portal, the entrance doors are overshadowed by a wrought-iron balcony, while all of

The Vanderbilt mansion's
main entrance.

The north and west porticoes
facing the river.

Plan of the Vanderbilt
mansion.

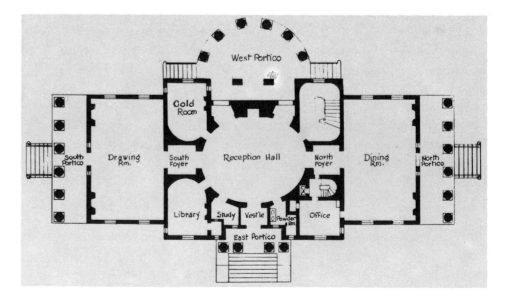

OPPOSITE: Mrs. Vanderbilt's bedroom.

the windows except those of the third floor are partially screened by decorative iron gratings.

Today the estate comprises 211 acres of open, parklike grounds, ranging from the mile-long natural terraces fronting the river to the flat, wide, tree-dotted lawns that sweep down to the highway from which the mansion itself is barely visible. Most of these neat, grassy areas, like the well-spaced trees (some of which are valuable imported specimens), are relics of earlier estates that occupied these acres. They are arranged with due regard to the meanderings of Crum Elbow Creek and the winding driveways and wagon roads that lead to the present mansion from the various outbuildings, staff cottages, and tall boundary gates, the whole inviting pleasance preparing the visitor for the air of quiet opulence that makes itself felt on entering the palatial residence.

In the huge oval reception hall, walls of green and white Italian marble vie for attention with the rich, cut-velvet hangings and a white-marble fireplace mantel with life-size caryatids which once adorned an Italian *palazzo*. Above this hangs a Flemish tapestry bearing the insignia of the Medici family of Renaissance times, while on either side of the fireplace—and elsewhere in the room—are antique Italian throne-like chairs upholstered in velvet and needlepoint, while a pair of French Renaissance cabinets in walnut inlaid with marble are disposed on either side of the front doorway.

This entire floor is of patterned mosaic in green, gray, and off-white. In its center stands a massive carved and gilded antique Italian table with a heavy porphyry top supporting a tall, 200-year-old clock in bronze and porcelain with a porphyry base. Interesting but somewhat less exciting are the male bust in Carrara marble on a Rococo-style wood stand, to the left of the fireplace, and its female counterpart at the other side of the hall.

Much of the reception hall's rich exuberance, however, is supplied by its architectural details, such as the twin pilasters of green marble flanking the doorways and flat-arched openings. These have capitals of white marble decorated with paterae, and above them is the green-tinted frieze, carved in an intricate anthemion design, with narrower decorative bands under the cornice. The single-paneled doors to the closed rooms, such as the library, study, and powder room, are enframed in eared surrounds of white marble, their horizontal pediments set off by decorative moldings above a paneled overdoor. Above these are rectangular slabs of green marble in off-white-marble crossetted frames bordered by egg-and-dart molding. On these walls a final exotic touch is provided by a series of triple-branched wall sconces of antique bronze whose frosted spherical globes look like so many giant pearls.

To one side of the reception hall is the drawing room which, in turn, opens onto the south portico. This drawing room runs the full length of the building and is one of the most elaborately decorated and furnished apartments in the mansion. Its

A stately bedroom reminiscent of French royalty.

OPPOSITE: The Gold Room.

The entrance hall.

paneling of Circassian walnut is carved in high relief in such areas as the arched overdoors, projecting at ceiling height to form a cornice, supported by carved modillions, which is carried right around the room. The overmantels are of plaster, and in the white-painted, coffered ceiling, some of the panels are filled with designs in gesso, while in others the decoration is confined to the borders.

At the south side of the room, twin fireplaces of gray Italian marble flank the wide portico doorway, while the end walls display seventeenth-century Florentine tapestries bearing the Medici arms. A pair of sixteenth-century Brussels tapestries illustrating scenes from the Trojan War hang astride the doorway to the reception hall.

With the exception of a Steinway grand piano, two Italian refectory tables, and a number of Chinese lamps, the drawing room furniture is French, mostly of the Louis XV period, the gilded chairs being covered in colorful tapestries, interspersed with a large canapé in gold brocade and an ormolu-mounted desk—a *bureau à cylindre* reminiscent of the famous Louis XV *bureau de roi*—adding their quota of interest to a magnificent ensemble.

A close second in the luxuriance of its interior is the 30 by 50-foot dining room to the north of the reception hall. Here the

206

OPPOSITE: The stairwell—replete with bronze and velvet.

A typical example of guest room furnishings.

Most of the guest bedrooms follow this pattern.

pièce de résistance is the magnificent coffered ceiling from an Italian villa, carved in minute detail and highlighted in gold. Its large central panel is painted with a celestial scene of clouds and cherubs, with smaller circular panels in geometric floral designs.

The walnut-paneled walls are also richly carved between fluted Ionic pilasters. All of this dark woodwork, together with the Burgundy-colored cut-velvet draperies, produces a somewhat somber effect relieved only by a pair of ceiling-high Italian Renaissance fireplace mantels in white marble and the reflective qualities of the magnificent 20 by 40-foot, 300-year-old multicolored Isfahan rug which occupies so much of the floor.

For the most part the furniture consists of reproductions of massive Louis XIV pieces, though several Florentine chairs share wall space with a fascinating pair of eighteenth-century planetaria of London make. Flanking the entrance from the tiny north foyer are two Ionic marble columns, duplicating those at the drawing room entrance, with a pair of seventeenth-century French tapestries, probably of Beauvais origin, hanging alongside them.

This dining furniture consists of a large, walnut-topped table with carved and painted legs, typical of the Louis XV era, which, when extended, could seat thirty people. Cross-stretchered chairs to match are upholstered in the same Burgundy cut-velvet as the draperies. Beyond this grouping stands a smaller table for use on less formal occasions.

Among the several smaller apartments opening off the reception hall is the Gold Room, which is almost overpowering in its lavish use of gilded decorative details and gold-colored fabrics, all amplified in intensity by the pair of tall pier glasses hung to face each other so that the reflections are repeated ad infinitum. Happily, in this miniature French salon, though the walls and doors are painted white, the ceiling is covered by an 1897 painting by the American artist Edward E. Simmons. Apart from its ormolu mounts and a pair of caryatids, the mantelpiece is of a red-veined marble.

On the other hand, the velvet draperies are in figured gold, as is the gilded screen with its stitchery panels and the upholstery of the twin, gilded *bergères*. Still more gold tones are supplied by a table with a black-marble top, the elaborate appliqués on walls and mantel, the equally ornate Rococo mantel clock, and the candelabra. Even the beautiful mantel urns of a blue granite finish are ormolu-trimmed, as is the Louis XV desk of inlaid tulipwood. Only the floor covering of pink velvet provides a background area of contrasting, positive color.

On the second floor of the mansion, the one room of peculiar and historic interest is the bedchamber formerly occupied by Mrs. Vanderbilt, reviving, as it does, the French monarchs' retiring and arising ceremonials of *le coucher* and *la levée*. This whole room attempts to reproduce the essentials of a queen's

bedroom in the Louis XV period style in all its lavish adornment.

Centering on the gold-embroidered, silk-canopied bed, with its spread of gold and pink taffeta, this area of the room is set off by a U-shaped curving rail at floor level and by a bow-fronted continuation of the room's frieze and cornice across the ceiling, the latter upheld by a white Corinthian column with a gilded capital on each corner pedestal of the front rail and pilasters against the side walls. Within this area the walls are covered in gold-tinted silk damask, while the double doors on either side of the bed arc surmounted by rich mythological paintings, gold-framed in Rococo style.

The rest of the walls are of gray-painted, gold-trimmed wooden panels between and around several pairs of doors, all topped by French paintings. There are two tall, gold-framed pier glasses whose crests consist of a flower-filled urn flanked by a pair of cupids in the round, clinging to floral swags. Beneath one of these huge mirrors is a marble-topped, Louis XV-style *bombé* commode decorated in bronze *doré*. In this area outside the bed enclosure the ceiling is dished, and the deep coved frieze is divided into figure-decorated white panels by curved modillions which support the molded gilt cornice.

Every piece of furniture in this room—commode, tables, chairs in brown-and-white cut velvet, and even the chaise longue in its modern gold slipcover—is an exact copy of a French original. Rich pieces in a rich setting, they are tied together by a silky, deep-piled beige rug adorned with huge floral medallions in pink, green, and tan, which was made especially for this room.

A corner of the lavishly decorated music room.

Antique French pieces dominate
the drawing room.

In contrast with his wife's room, Frederick Vanderbilt's
bedchamber is almost baronial—and very masculine—in its
massive detail, its emphasis on crimson and gold (common to
the deep, overstuffed pieces), its finely modeled, heavily
timbered coved ceiling with its twin ram's-head brackets, and
the almost barbaric twisted Baroque columns garlanded with
golden flowers. The primitive, Garden-of-Eden-patterned,
seventeenth-century Flemish tapestries with their profusion of
gnarled trees and lush fruits are ruthlessly cut apart to permit
of door openings in the walls to which they are firmly attached.

Circassian walnut forms the room's paneled dado, as well as
the attached columns and wooden canopy of the great bed,
its ornate side panels, and the bed ends which are deeply
carved like a medieval chest, with the addition of terminal
caryatids. The figure-ornamented dresser also forms a part of
the general woodwork. The bed's canopy backboard is
covered in red and gold velvet, incorporating a ducal crown
in its central design—a motif reappearing in the back upholstery of a seventeenth-century *os de mouton* settee, companion
piece to an equally early tall, scroll-back chair, both of which
are upholstered in a Savonnerie tapestry, whose pattern and
coloring have something in common with those of the Turkish
rugs gracing the floor.

These two bedrooms differ considerably from the guest rooms, particularly the Blue Room, which gets its name from the background color of the European wallpaper which has an allover floral pattern in white. As usual in these transient bedchambers, the floor is covered with fitted, short-pile carpeting, augmented in this instance by a Ghiordes prayer rug in blue, red, and beige, fronting the hearth.

All of the trim in this room, including the door and its elaborate framing, the wooden window lambrequins, the extra-tall baseboard, and the frieze and cornice, is painted white to match the ceiling. Against this predominantly white and blue background, the eighteenth-century-style French furniture pieces are more or less unified, the cane-back bed being painted gray and gold under its blue and gold canopy of silk damask, with which the pale-blue lace spread embroidered in tan silk is in happy contrast. The window draperies with their voluminous swags and cascades are of gold lined with blue, while the upholstered pieces are covered in either blue or gold cotton damask to introduce a note of variety into an overall combination of subdued colors and accents that is gay but restful and likely to appeal to its occasional occupants.

The rest of the rooms, such as the library, the study, the Empire Room, and those other guest quarters identified by their color schemes, follow the same furnishing plan of using the finest pieces available, whether antiques, perfect copies, or modern, to produce interiors that combine comfort with charm and utility with aesthetic satisfaction. This great house, once among the finest of country houses designed for luxurious living, on occasion serving as a *maison de plaisance*, is now, along with its extensive grounds, a National Historic Site.

The Blue Room—one of those set apart for guests.

The main dining room with its antique Italian ceiling.

Natchez, Mississippi

CHEROKEE

Spanish-American Classical

The high terrace entrance to
Cherokee.

The Natchez mansion, which has been known as Cherokee for
the past 158 years, is as fascinating as its name, largely because
of its contrasting front and rear faces which recall two anti-
thetical periods in the nation's history and reveal the reason for
the unusual interior arrangement that has resulted.

To understand its origins one must delve briefly into history.
The first house on this site—a long, one-room-wide structure—
was erected in 1794 by a certain Jesse Greenfield who had
acquired a land grant from the Spanish authorities in whose
territory it then lay. In 1799, a year after the area was ceded
to the United States, the property was sold to a Scot named
Robert Dunbar who lived in the house 10 years before turning
it over to a new owner, David Michie. It was this David
Michie who, with the help of an architect named Levi G.
Weeks, added the classical front in that same year (1810),
together with the five additional rooms immediately behind it.
This entire unified structure thus became the great house that
is known to this day as Cherokee. Today it is the home of
Mr. and Mrs. Charles J. Byrne who restored it to its original
condition while adding the amenities that make it not only
suitable for modern living but a joy to inhabit.

Perched on a steep, roadside hillock, above tall brick and
concrete retaining walls, the mansion is approached by steep
stone steps that terminate in a flower-banked brick terrace,
above which the dead-white classical façade is bathed in the
light of the afternoon sun. That façade has all the austere air
of a Greek temple *in antis* with its twin columns and pilasters
and its pedimented portico. Only the short, parapeted side
extensions are out of character, yet they, too, add to the total
effect, giving it a substance and a solidity it otherwise would
not have. In this astringently beautiful façade, only the demi-
lune window in the pediment and the rectangular ones
flanking the deep central portico suggest an early nineteenth-
century habitation, in bold contrast to the original house which
is still pure eighteenth-century Spanish.

OPPOSITE TOP: The Cherokee rear lawn.
OPPOSITE BOTTOM: The dining room looking into
the breakfast room.

In plan this house is T-shaped, thanks to the addition of a series of rooms between the principal façade and the original front of the longer Spanish house behind it. The gable-ended Spanish section, still intact except for some interior rearrangement, is taller at the back because of the sloping land which exposes much of the rear wall of the basement. This apparent extra height adds considerably to the dignity of the whole structure, especially as seen from the back.

On the main floor, the original unit contributes two small bedrooms and one master bedroom and associated bathrooms, together with a huge "back" drawing room, a kitchen, and a highly intriguing pair of longitudinal halls, each of which extends from one side of the central drawing room to the nearest gable end. This quite unusual arrangement banishes the principal staircase—a lovely "winder"—to the end of one of these long halls and necessitates the location of the basement stair in a corner of the back dining room. Otherwise there is considerable economy of space, with a nice reception hall opening off the portico, which gives access to the back drawing room from which the halls are reached. Obviously this arrangement adds considerably to the size of that drawing room, though demanding traffic space to the halls.

On the opposite side of the house is another bedroom suite and upstairs a 45-foot-long center hall, off which open three bedrooms and a sitting room. Here the ceilings are much lower. The front block of rooms is equally compact, with a small drawing room, or parlor, to one side of the almost-square reception hall and a dining room on the other, while the portico itself is flanked by two tiny rooms, one a study, the other a simple but charming breakfast room. Other bedrooms are located on the second, attic, floor and in the servants' quarters in the raised basement, where they have ample window area and a very pleasant outlook over the terrace gardens.

Returning to the main section, the portico is a delightful feature though quite plain and formal—and massive. Its ceiling-high main entrance has paneled pilasters and a molded, flat-topped overdoor. A tall sash window on either side of it, complete with louvered blinds, occupies the rest of the front wall space inside the portico. There is, however, a doorway opening from the portico into each of the small end rooms, breaking up the blank areas of the end walls. From the center of the portico ceiling hangs a tapering lantern in antique brass decorated with filigree—a touch of the exquisite in a classical setting.

From either end, the deep, central block of the mansion gains in stature from the tall, parapeted gables which incorporate the twin end chimneys, though they are largely shielded from view by towering trees. At the rear, the lovely formal gardens, with their wide stone steps and tall box hedges, have as background the two-story, white stuccoed walls, with the huge central drawing room windows, and ten smaller ones on

OPPOSITE: The Cherokee back drawing room.

The brick-paved dining terrace.

A glimpse of the house from
the rear garden.

Cherokee's end stair hall.

two levels; plus wall and walk lanterns opening at one end onto a wide, brick-paved dining terrace with its white-painted, filigreed, iron and plate-glass furniture. The terrace terminates in a gabled, cottagelike potting shed that is every bit as ornamental as it is useful. On all sides are deep box borders, while much of the paved area lies in the shade of a tall, wide-spreading tree.

From inside, the two great windows of the large drawing room look out upon another formal garden centered on a marble fountain, whose ground-level basin is filled with blossoms according to season, as are the four, brick-bordered corner plots, shaped like the spandrels of an arch, that give shape to the ring of grass around it, all against a background of heavy foliage. This, then, is the logical room from which to begin inspection of the vastly interesting interiors whose architectural features do so much to establish the character of the furnished rooms.

White and pale green are the key colors of the back drawing room as determined by the trim, ceiling, and walls respectively. And there is a great deal of trim here with the deep molded cornice and baseboard, the wide doorframes, reveal panels, the frames and mountings of the huge overdoor lights and sidelights, all quite angular, following the pattern of the front door from the reception hall to the portico. The choice of a rug for this "chopped-up" room, therefore, was of the greatest importance, since part of its function would be to pick up the basic colors in such a way as to diffuse the general pattern and scatter the highlights to provide a feeling of uniformity without increasing the impact of these great masses of colorless reflective surfaces. This was achieved by covering the entire floor with Brussels carpeting having a gray-green ground with a widely spread pattern of medallions in darker green and off-white, with occasional tangerine centers. On it two English armchairs, one in green leather and the other in cut velvet in muted tones, are placed near the windows, counterbalanced by a Sheraton sofa in off-white linen damask, which, together with a huge floral display in a Staffordshire bowl of French design on a table behind it, diverts traffic from the central hall door.

Nearby, a tall mahogany *étagère* and a grand piano with its long upholstered stool reduce the impact of adjoining white doors while, at the opposite side of the room, a Regency chair in coral damask and a low chest distract attention from the railing of the basement stairs.

Other pieces include a Sheraton writing desk with a galleried top, two Late Empire chairs in mahogany, and an exquisite Moorish octagonal taboret inlaid with ivory, leaving ample room for perambulating to and from any of the inviting rooms whose doors seem always open to provide interesting vistas. The magnificent crystal chandelier in the middle of the room's ceiling hangs high enough to illuminate the whole room without posing any hazard to those walking underneath.

A quite different decorating situation is presented by the

parlor (small drawing room), which has a much more formal atmosphere—and less of a family room or music room air. These light-gray walls are bathed in daylight filtered through sheer curtains of Brussels lace and translucent draperies of silver brocade. The result is a pleasantly neutral background, both for the tufted, pale-pink upholstery of the rosewood, chair-back-ended "sociable" of 1845, and its accompanying armchairs, and the much more restrained sofa and balloon-back side chairs facing them across the pink-veined, black-marble fireplace. Uniting the two groups is a splendid Aubusson rug, whose pale-beige ground is enlivened by a floral design in muted colors.

Touches of gold are provided by the Rococo frame of the huge French overmantel mirror and the much taller pier glass, also French, on a low, bracketed shelf that graces the side wall. Reflected in both is a fine French chandelier of many-faceted teardrop crystals, and on the overmantel one also glimpses a vitrine opposite the fireplace, with its display of fine glassware. Also reflected in it is the entire mantel garniture, all French in provenance, consisting of tall, gilt candelabra, a clock enclosed in a gold and crystal case, and a pair of decorated Meissen vases. These display pink blossoms—a touch of color repeated in a gold-framed portrait over the pink sofa, in twin porcelain lamp bases on a tripod table, and a Chippendale card table against the same wall, as well as in the rug, thus tying the whole room together into a delicate and somewhat feminine whole.

The Victorian parlor.

The tiny but attractive breakfast room.

The master bedroom.

On the other side of the entrance hall, the dining room strikes an entirely different note, even though the overmantel mirror is a duplicate of that in the parlor. For one thing, the fireplace is white to match the door and window trim, the walls being covered with a light champagne silk which offers but little contrast to the pale-beige ground of the rug, which was made in Hong Kong especially for this room. This scheme of using analogous tints and shades of the same basic color, even in the lighter tones, has given the room a sophisticated look without a suggestion of the dramatic, creating an enchanting background for the beautiful Sheraton mahogany furniture.

Over the Brussels-lace glass curtains are side draperies in pale gold satin, under fringed pelmets in a slightly darker champagne color. The light rug has a wide border in a soft green with touches of autumn tones, widened in front of the fireplace to substitute for a hearth rug and incorporating at its center—under the dining table—a large oval medallion in delicate greens, orange, and gray. The late Sheraton chairs of klismos form with rope-turned crestings are covered in a pale-green wool brocade several shades lighter than the green in the rug on which they stand. Here, too, gold accents are provided by the frames of the overmantel mirror—a twin to that of the parlor—and several pictures including an especially fine English landscape of the Richard Wilson school over the bow-fronted sideboard.

Through the open door of this room can be seen the cozy round-table grouping in the adjoining breakfast room, against the cheerful background of a white wallpaper patterned in a brown, beige, copper, and green looping stencil design enframing a bird on a twig, in the same colors, endlessly repeated. In strong contrast is the cascading pelmet, half the length of the window, in a soft gray-green with a green and gold fringe, the green being repeated in a deeper tone by the fitted broadloom carpet.

Centered in the ceiling is a globe-type hanging lantern designed to illuminate the circular pedestal table beneath it. Here also is an English tilt-top tea table with a deep apron in beautifully grained mahogany, surrounded by chairs of the same period covered in a checkered cotton damask of a grayish green on white. In a corner by the window is a spider-legged, three-decker whatnot, used when required as a serving table in this tiny room, otherwise doing duty as a stand for a tiny crystal-bedecked astral lamp (now electrified), and displaying colorful antique glass dishes on the lower shelves.

The similar small room at the other side of the portico is equipped for use as a study, though from its exotic equipment it could equally well serve the needs of an invalid—or recluse. Nevertheless it is quite cheerful, with walls punctuated by Audubon bird prints in gold frames. The walls are covered with a Japanese grass paper of a rich celadon green, while the

The back drawing room with its exit to the garden.

The light and airy dining room.

The study with its gout chair
and other antiques.

floor-length draperies are of a green and beige documentary print. Other color is provided by the blue-and-gold-striped damask covering a wing chair and the green, cream, and red-striped satin seat of a Phyfe open-arm chair. The largest piece of furniture is an English Sheraton secretary-bookcase of mahogany with a line inlay and an oval patera on the drop front. The most interesting item, however, is the old-time gout chair with its sliding leg rest, as useful today as any "lolling" chair. Heavily constructed of mahogany, it is upholstered in green leather. Next to it is another curiosity in the form of an antique night commode disguised as a galleried lampstand with what appear to be a small cupboard and a drawer.

Finally, in the Spanish section of the mansion is the comparatively quite large master bedroom, which is equipped with a fireplace having a substantial white-painted mantel, a relic of territorial days. In this room pink again comes into its own, the wall being of a pinkish tint sprinkled with roses of a deeper hue, relieved by creeperlike green foliage. The draperies likewise are of a pale-pink silk over white glass curtains. Happily, ceiling and trim are white, and, in contrast, the modern carpeting is of a dark green, though even this harbors a large open-arm chair in a soft-pink velvet that covers seat, back, and arms.

Flanking one of the windows are two heavy, mahogany tester beds, made locally, with testers, dusters, and spreads all in white, though the small, beautifully shaped sofa at the foot of one of them is upholstered in a silvery gray.

Under the window between the two tall beds is an oval mahogany table with a heavy, bulbous, four-footed pedestal, while by the side of another window is a rather graceful swinging dressing glass on casters, contending for space with a huge mahogany chest-on-chest—nine drawers in all—flat-topped, with fluted, canted corners and bracket feet. This whole array of massive pieces is illuminated by a delightful six-branch chandelier of gleaming crystal—a gracious touch in a room that otherwise concentrates on solid comfort as seen through Victorian eyes.

Reidsville, North Carolina

CHINQUA-PENN PLANTATION HOUSE

Gracious Living in a Jewel Box

Chinqua-Penn is a delightful oddity, from its name to the Chinese temple that guards its swimming pool and the fabulous contents of the mansion itself. That name actually is a play on words—the cognomen "Penn" grafted onto that of the chinqua-pin (the dwarf chestnut that once throve on these hills)—an eponym in reverse!

The mansion itself is a low, rambling, Y-shaped structure set amid the loveliest of gardens on a wooded knoll; inside it has all the air of an oversized jewel casket, as crammed with fantastic treasures as Aladdin's cave—and a thousand times more real.

This twenty-seven-room residence was put together from local stone and hewn logs, with the help of an imported Italian mason, over a period of two years from 1923 to 1925. The bifurcated plan is a natural consequence of condensing the two-story house to a minimum overall length while endowing most of the rooms with a view in two directions and trapping sunlight the whole day long.

The house follows no particular architectural style with its low-pitched roof, deep eaves, oddly placed chimneys, and occasional square, hip-roofed dormers. These, happily, are in keeping with the rugged, colorful, random stone walls which give an impression of impregnable strength that the huge, grillelike entrance door, salvaged from a Venetian *palazzo,* and the heavily timbered oriel window above it do nothing to dispel. The overall aspect is that of a baronial mansion with the lines of a massive, oversized country cottage having certain exotic touches—such as a gable end formed of three heavily mullioned windows extending from sills to plates—that stimulate the imagination and arouse one's curiosity in no uncertain manner.

As befits its intriguing style, the house is set within fabulous gardens mixing ivy-clad walls with boxwood mounds and trimmed hedges, high and low; dense clumps of tall, flowering trees; occasional rose and herb gardens; and marble fountains

The rose garden terrace.

221

The inviting breakfast room
in gay colors.

dotting the winding walks and low-hedged *allées*, culminating
in the aforementioned temple and pool. For the former the
Penns brought from China the lanterns, bells, terra-cotta
figures, and other furnishings, such as the stone gargoyles and
ancient roof tiles that complete the illusion of timelessness. The
building itself was an authentic copy made in this country but
painted in the true shades of red and blue, with black, gold,
and green decorative details, by a Chinese artist. Outside it
stands a great stone Buddha that once graced the Chinese
garden of Mrs. Penn's mother in Niagara Falls.

From the temple a treelined *allée* climbs the slope toward
the house, terminating in a boxwood circle surrounding a
lovely bronze fountain from the gardens of Versailles. At the
other side of the house, the rose garden centers on another
white-marble fountain, separated by a wing from the long
formal garden and parterre, and all through these planted

OPPOSITE TOP: A Chinese temple guards the swimming pool.

OPPOSITE BOTTOM: On the foyer's floor is a fabulous
Karabagh Caucasian rug.

areas are scattered huge earthenware jars, stone statues, and benches, intensifying the old-time atmosphere redolent with nostalgia for things long gone.

In the front hall, or foyer, of the mansion that feeling is intensified as one moves among recognizable rarities not often encountered and strange antiquities, all in a natural setting that eliminates any museumlike atmosphere. This is not furnishing as it is ordinarily conceived, for practically everything is either beautiful or rare, or both, in this most livable of homes. On the other hand, after seeing the pair of modern portraits on the foyer walls, one would not be too astonished to come face-to-face with Jefferson or Beatrice Penn. It was the latter, then a widow, who presented the estate to the University of North Carolina in 1959.

Between the foyer's walls of sunken oak paneling, with its stop-fluted Ionic pilasters and gouge-carved cornice—relic of an English manor house of yesterday—stretches the deep tray ceiling, a pattern of squares, diamonds, and sweeping curves of gesso molding, displaying flower sprays at the corners and rosettes in the panel centers, and all tinted like ancient parchment. Facing it on the floor below is a fabulous Karabagh Caucasian rug in pinks, blues, and pale gold, narrow enough to let the medley of exotic furniture pieces stand on the dark parquet floor around it.

Undoubtedly the most colorful furniture piece in this hall is the gold and vermilion Chinese lacquered cabinet whose large, square bronze latch is finely engraved. On the other hand, more recently made but by far the most historic in its associations is a reproduction in ivory-inlaid teak of the "golden" chair of Egypt's famous king Tutankhamen of 1358 B.C.

Somewhat more recent, yet equally surprising, is the mixed group of artifacts consisting of a Chinese temple table from Peking, ornately carved and inlaid with ivory, over which hangs a Nepalese altar piece in its decorative gold frame, flanked by exquisite boxlike bracketed shelves, on which rest a pair of colorful gilt-bronze and silver wine pourers from Nepal, studded with semiprecious stones. In close proximity to these are terra-cotta statuettes from a Peking tomb, a Chinese ivory-inlaid tea table, and temple lamps with translucent shades made from animal horns softened in boiling water, split, and flattened for that purpose. An eclectic touch is provided by an English cabinet of the Jacobean period and a square-backed chair dating from the reign of Charles II, *circa* 1665, upholstered in Turkey work.

Off one side of the foyer, facing this chair, is an opening through which one walks into a semihexagonal, tile-floored area and down three tile-faced stone steps to the floor of the main living room. Based on the middle step are four sculptured stone columns in the Italian Renaissance style, which support three semicircular arches with paneled soffits, their urn-adorned tops forming a decorative parapet serving as balustrade to a balcony above.

A magnificent tapestry
dominates the main living room.

225

OPPOSITE: The living room in gold-colored plaster.

This living room is the most spacious apartment in the house, towering up into the heavily timbered open roof. All the lower timbers are boxed and painted in vividly colorful designs, each curved wall bracket terminating in a huge carved and painted pendant. From the central beam, which runs the length of the room, hang large octagonal lanterns of decorative metalwork from a Chinese temple. At one side, over a tall window, is an open iron-railed gallery, and above the next one is a *bombé*-shaped wrought-iron balcony. All these windows are draped in silver brocaded damask.

In this area the thick stone wall is covered with gold-colored plaster, but at the ends of the room the multicolored stonework is exposed, providing a most effective contrast. Two of the upper windows flanking the arched entrance are given a medieval—or Spanish—touch by a pair of triangular rugs hung from the sills—in the manner of boxes at the *corrida!* Elsewhere, lower sections of the walls are covered by pecky-cypress paneling, which not only adds a note of sumptuousness but, on either side of the archway, provides a fitting background for a pair of faïence hypocausts distributing warm air to this baronial hall. These twin structures are shaped like fireplace mantels—minus the central fireplace opening—with end pilasters and bracketed cornice, all composed of brilliantly colored tiles in dark blue, white, pale green, and gold. The top, which forms a shallow shelf, is all blue, and centered on

The library, rich in detail.

each is a tall terra-cotta statuette. These are Chinese, of the Tang Dynasty. One depicts the Spirit of the Earth; the other the Spirit of the Moon, and each is flanked by a pair of Limoges porcelain panels of saints against the cypress-wood background.

At the other end of the room is a sculptured stone fireplace of Florentine design taken from an Italian nobleman's *palazzo*, and over it hangs a 16-foot Brussels tapestry 300 years old. Across the opening is an elaborately carved fire screen, faced by a fireside bench covered in a rich green velvet. On the polished oak floor of this room are large velvet rugs in a pale green, with a clipped border design in a silvery tone—colors that harmonize beautifully with the *voyeuse*, the X-stool, and the banquettes in rose velvet, as well as with the entrancing colors of a Beauvais tapestry on one sofa, the silver brocade of another, and the cream-colored silk damask of a third.

Behind the cream-colored sofa is a carved table of sixteenth-century Italian design. This has a brown-marble top on which are twin lamps with Italian Renaissance-style bronze bases and quite unusual tasseled shades in a gold fabric patterned in a pale green. Between these is an exquisite silver bowl of Egyptian origin filled with flowers—a bouquet that contrasts happily with an enormous jardiniere of Carrara marble on the floor nearby, from which springs a veritable fountain of variegated foliage, and a highly dissimilar carved-stone, urn-topped pedestal sprouting spiky leaves of palm and a cascade of English ivy.

Among the remaining notable pieces here are a pair of

A short hallway known
as the solarium.

Venetian commodes of 1780 or so, over which tall mirrors incorporate an oval painting in the upper panel. There is also a third commode—a fine Louis XVI type, elaborately bedecked with ormolu mountings beneath a brown-veined marble top.

Flanking one of the swell-front Venetian pieces is a pair of finely carved, gilded, and painted *torchères* serving as floor lamps, with large pink shades, while across the room is another pair of *torchère*-based lamps separated by a long, oval table of a Venetian type, which has a finely carved border to its one-piece top.

There are many other rooms in this mansion, some of them quite small, which are the ultimate expression of the exquisite in furnishing and background design. They display within their four walls unified groupings of world treasures, often the finest of their kind, one short peep into any one of them being sufficient to reveal their total beauty and interest and the artistic details that made their grouping together inevitable.

One such private apartment in which are assembled the treasures most dear to the Penns for what they represented in the way of artistic concept and execution rather than their monetary value is the private sitting room. Here the walls, floor, and ceiling are designed to form a background against which the contents will meld into one pleasing whole without any loss of individual identity. The canvas-covered walls, for example, painted a bluish green, are framed between an intricately carved and heavy cornice, delicately detailed, and a dado of hexagonal panels with a narrow top rail carved in a riband and daisy design and a reel beading at the base, all painted a contrasting grayish green. This dado is interrupted only by the elaborately carved entrance doorway to the room and that of a corner cupboard, with their overdoors and bracketed pediments leaving little space unadorned. The ceiling is completely plain, but the floor is wholly covered by a modern short-pile rug in beige with a ribbonlike sculptured pattern. The floor-length draperies are of gold silk damask with pink and gold tassels.

The largest and most impressive unit in this jewel box is the fireplace and its overmantel—a Roman antique with overall decoration. The lower section, standing on a hearth of polished black and white marble, is inlaid with a dozen large beige and blue tiles, depicting in low relief male and female workers busy with a variety of projects. The heavy mantelshelf is elaborately carved and topped by a slab of white marble, while above it the overmantel, with its overall carved and incised decoration, forms a twin-pilastered frame for a circular painting of plump, winged *cherubini* draping the clouds with a garland of flowers. On the mantelshelf is a 1770 silver chalice from Italy, supported by a draped female figure, while on the wall nearby are hung Russian icons and an Italian crucifix of fascinating design.

An especially massive work of art is a tall, carved piece

OPPOSITE BOTTOM: The owner's personal living room displays special treasures.

in wood consisting of a sturdy sixteenth-century Spanish vargueno, with a four-door cabinet as base. The front of this vargueno—reputedly made for some peripatetic bishop—is normally dropped to reveal the ivory-pillared niches and the porcelain panels in the drawer fronts depicting scenes from the life of the Nazarene, plus the elaborately paneled doors over which hangs the primitive but effective locking device. With this twin piece goes a simple Spanish country chair, plain in line but interesting because of its simple leather seat, which is nailed only to the side members to give it a comfortable sag, as is the similar, unpadded leather back strip.

This chair might very well be compared with the sixteenth-century Spanish bench in walnut, which is a feature of the short hallway known as the solarium. Their frames and arm-rests are similar in construction, but the bench has a padded seat covered in velvet edged with a fringed braid. The back, however, is of embroidered, heavily fringed fabric supported only at the ends by upward extensions of the rear leg posts.

The doorframes at either end of this hallway are amazing confections of the Italian Renaissance period, the attached columns carved with flowering and fruiting vines, the door stiles and top rail a mass of beribboned fruit and blossoms. The architrave is made up of seven recessed panels decorated with medallions, and even the intervening brackets supporting the cornice have carved masks on their lower faces. Above these the heavy, dentiled cornice agreeably relates to notched and painted joists in red and white.

The incredible vargueno in the sitting room.

The Penns' dining room.

Both the walls and the floor are of travertine, the latter inlaid with squares and large hollow diamonds of a dark blue. Sculptured out of Carrara marble are the console tables, floor-lamp bases, and sixteenth-century flower fonts with their multiple, carved colonnettes, every one a delight to the eye. On one of the consoles is a cast-bronze head of Buddha, a relic of fifteenth-century China.

On these walls are pictures from old China—including two rare sand paintings—which hobnob with others from India and Italy, all of which contrast interestingly with a Queen Anne wing chair and side chairs of Spanish provenance, the common denominator being a mutual antiquity.

In Chinqua-Penn's library the same incredible richness of detail is everywhere evident, quite apart from the superb bindings of many of the books. Vivid Spanish tiles in plain and floral designs line the alcove, and all of the structural woodwork is decorated in some degree—the ceiling joists slash-carved (and apparently once painted over); the deep cornice and casings of the bookshelves, set against plastered walls, are painted green, with the lower sections concealed behind carved panels.

Over the recessed windows are draperies of heavy cotton

brocade in a deep, rich red, with wide borders in gold, adding to the luxurious feel of the whole room.

The armchairs here are large and formal sixteenth- and seventeenth-century high-back Italian designs in red velvet with a gold fringe, plus a huge seventeenth-century Persian armchair inlaid with ivory, whose seat and back are upholstered in tooled leather. There is also an extravagantly styled seventeenth-century Italian taboret with a round, framed Loredo-Chiano marble top and three scrolled legs of carved and gilded wood, with the apron and stretchers in the leaf-chain pattern.

At one end of the room is a Louis XV open-armed canapé in silk damask of an oversized pattern in keeping with the huge splashes of gold embroidery on the Spanish chair backs. Through the open doorway of this room can be seen a seventeenth-century open-fronted and backless display rack of five shelves whose paneled pilasters, which support a broken pediment, are covered with silk in a pattern of vases and plants. On each of the adjustable shelves is an almost priceless example of antique Chinese porcelain.

Elsewhere in this library there are miscellaneous treasures ranging from a Florentine sofa in petit point to Italian tables and bronze floor lamps, all adding to the seemingly endless collection of smaller curiosities representing the arts of the ages which few could not benefit from by examining at close quarters.

A bedroom in the French Empire style.

The Italian master bedroom.

ELEUTHERIAN MILLS

From Paris to the Brandywine

The rear balcony looking
toward the deep woods.

Of all existing houses sufficiently impressive in size and quality to be called a mansion, or even great house, few are so utterly homelike in as delightful a manner as the English Georgian designed by one Frenchman for another and dubbed Eleutherian Mills (Anglice: Liberty Mills) in recognition of the family's deliverance from the excesses of the French Revolution toward the end of the eighteenth century. Apparently, however, that comfortable, livable quality, tempered with an air of restrained elegance within and dignified charm without was ever a characteristic of this mansion during its occupation by E. I. Du Pont and his descendants for more than 150 years. A glance at it today may well reveal the reasons why. First, though, it is necessary to delve a little into history.

This pleasantly dignified yet unpretentious mansion actually began as a somewhat smaller house built in 1803 on a wooded ridge overlooking the Brandywine River which transverses the sixty-five-acre tract bought by Du Pont as the site of a powder mill he was to establish. Plans for the house had been drawn by Du Pont's associate, Peter Bauduy, a French refugee from Santo Domingo, whose son later married into the family. These called for a two-and-a-half-story structure of local stone with a stucco finish, and this was completed within a year.

In 1853, a lean-to was removed from one gable to permit the addition of a substantial wing at each end. These wings not only supplied the needed extra space for an expanding family but also balanced the central structure and gave the whole house an added importance. Much later the front portico was also removed to the rear piazza to replace the original colonnaded one, onto which the rear hallway door still opens, leaving a central pedimented doorway as the main entrance.

The house, as it now exists, therefore consists of a central block with parapeted, twin end chimneys and three tall dormers on each slope of the main roof. At either end is a short, two-story extension, one room wide, which is set back both front and rear. Each has a dormer on both roof slopes, but all fire-

Eleutherian Mills' main façade.

places are connected into the original chimneys. Since the land falls away sharply beyond each gable, iron-railed terraces were constructed, giving access to the ground level. Below the house the hillsides are also terraced and planted, the hedges being trimmed into geometrical shapes to disguise the stone ruins of older buildings and outline the stone steps. Still existing at the bottom of this slope is the old road which follows the course of the river, flanked by the picturesque ruins of stone buildings, widely spaced, which formed the original powder mill. The proximity of the house to these buildings is a reminder that, following French tradition, the management undertook to share the risks of explosions with their workers. Twice the house was wrecked, and several times it was damaged as a result of this propinquity. Today, however, the house floor plans differ little from those drawn in the late 1850's, though some of the rooms have newer uses.

One reason for this mansion's enviable lived-in atmosphere, and one not deliberately planned, is that the interiors are far from palatial, the rooms being designed for comfort rather than to impress. Moreover they are furnished with predominantly American pieces, ranging in style from those of the Pilgrim era to the Federal period, since most of the family-related pieces had become dispersed.

An 1852 sketch of the house by Henry A. Du Pont.

233

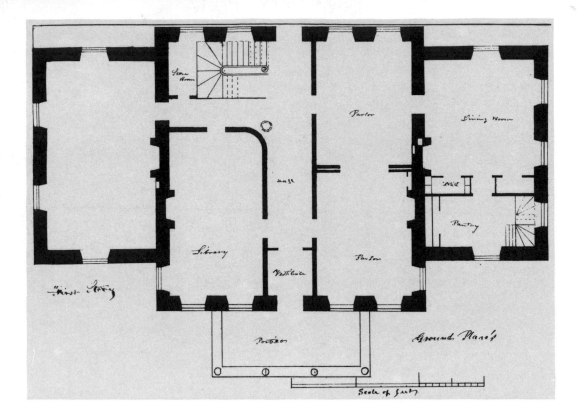

An early plan of the proposed house.

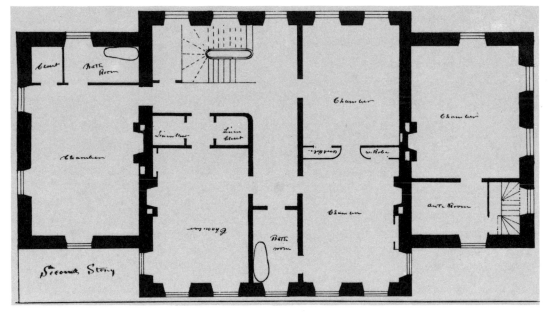

The second story as originally planned.

The most attractive feature of the house today is the spacious through hallway, with its wide, winding stairway, alongside which the twin rear doors, glass-paned from top to bottom and fanlighted above, open onto the lovely rear balcony. There, among the treetops, it is possible to look down upon the topiary-bordered gardens below. The stair itself, with its sweeping handrails, its delicate, white-painted balusters, its curving lower steps and slender newel-posts, is a work of art. Close by, the twin glass doors are flanked by tall, slender, mahogany tripod stands, each supporting a delicate glass vase in mandarin red. The walls, ceiling, and stair risers are a pale green, the cornices deeply dentiled, the walls paneled, the trim finely molded, and the half-dozen sturdy mahogany doors with raised panels are richly bold in contrast. Here also is a mahogany, serpentine-fronted cabinet and a Sheraton-style side chair dating from the early nineteenth century. The only color is supplied by small hooked throw rugs in a variety of patterns, but the total effect, day or night, is one of glowing charm.

In making over the interiors, one major operation was the turning of the twin parlors into the present drawing room by removing the sliding doors between them. This is now the largest room in the house and somewhat attenuated in appearance thanks to the twin fireplaces and the multiple doors of dark mahogany that face them, all tending to draw the side walls closer visually. Happily that fault is largely nullified by the clever grouping of the furniture pieces.

Today the drawing room furniture is a mixture of styles, representing the tastes of several generations, made to seem more homogeneous by the surrounding pale-pink walls, a color which sets off the pale green of windows and doors and presents them as separate units in the same manner as the larger movable units. The twin fireplaces are, of course, natural centers of attention, their pale-green painted wood surfaces decorated with gouge and punch work, which is emphasized by their marble facings. Nevertheless the colors and contours of the furniture have been well blended, the groupings constituting a counterattraction, and both benefit from the contrast, particularly since the walls are not at all overcrowded. On the floor the small but lively rugs, including an Aubusson in pink

The mansion today in its magnificent setting.

An 1840 oil painting of the house and powder mills.

In the morning room two pine corner boards display Export china.

and gold, add scattered highlights, so that the total effect is neither haphazard nor strictly formal.

The most interesting piece in this room is undoubtedly the fortuitous but happy combination of a Hepplewhite bookcase with brass finials and an oxbow-fronted Chippendale desk which seems to have been made to support it. This is flanked by two mahogany doors facing a recess between the twin fireplaces on the opposite wall. An equally commendable arrangement is the centering of a desk on each end wall of the room, one of them a Hepplewhite tambour-style piece, inlaid and banded, the other a transitional-type mahogany desk with a drop front, dating from about 1800. Against the walls there are several pieces of American-made furniture, such as a small tambour desk, a writing table, and a sewing table in the style of Phyfe.

On the wall there are five major paintings of various members of the family, including Pierre Samuel Du Pont and Colonel Henry A. Du Pont and his wife, plus an English painting of the 1830 period depicting Venus and Neptune.

As to the remaining furniture, the chairs represent a *mélange* of styles from Hepplewhite shield backs to Martha Washington "lolling" chairs and a few square-back side chairs, probably "after" Sheraton, all interspersed with little tripod tables in a variety of designs.

Of equal interest is the dining room, which is both unified and visually expanded by a colorful block paper covering the walls from the cream-painted dado to the handsome dentiled cornice. Though printed in France, this paper depicts an American historic scene, with coaches and human figures in costume of the 1830's. This and a white-marble facing set off to advantage the tall, wooden, mahogany-colored mantel adorned with swags and urns in gesso, the pilasters and borders fluted and punched. On the mantel is a colorful arrangement of unglazed Chinese burial figurines representing a variety of trades.

The three-part, round-ended dining table, with only the two drop-leaf end sections in use, is Hepplewhite of around 1800 with line inlay. Used with it are eight Hepplewhite shield-back chairs with five banisters and fluted legs, only one of which, curiously enough, has spade feet. Two other intruders are a pair of five-sheaf-style chairs with hollow seats, and stretchers, which actually maintain a sense of uniformity despite their individual differences, undoubtedly because of the matching seat fabric of pale-green silk brocade.

Besides a small, low-backed serving table in a corner by the unusual mahogany double doors—one opening in, the other opening out—this room contains a Hepplewhite mahogany sideboard. Scattered around the otherwise bare floor—a natural phenomenon in the early 1800's—are five gaily-hued hooked rugs, mostly in deep reds and blues on a light-beige ground.

Somewhat less formally furnished—and finished—than the preceding is the morning room. It is also much smaller and more cozy, with its light-colored wallpaper in a neat, checkered design, emphasizing the dark-stained pine of the cornice, window trim, dado, and fireplace, all of which closely match the oak floor in color. The top rail of the dado is ornamented with gouge work—a pattern of swags and tassels that is carried across the lower border molding of the mantelpiece. The mantel itself is even more highly decorated with gouge and chisel carving, including weeping willows over the pilasters and a decorative panel between them.

Two low, New England-style corner cupboards in pine with shell moldings are built in. These serve to display pieces of Oriental Export china of the 1780 to 1810 period, while on the wall between the cupboards is a group of nineteenth-century locomotive prints. Below them stands an unusual trestle table (*circa* 1850) with arched feet, rounded corners, and no stretcher.

Other details include a maple slant-top desk of 1770 and a Chippendale-style chair.

The morning room fireplace mantel is gouge-and-chisel-carved.

These feet, incidentally, suggest the influence of the French Comisards or the Shaker sect. There is also a neat, but simple, slant-top desk in maple with bracket feet (*circa* 1770), paired with a maple chair in the Chippendale straight-legged style. This is thought to have been made in Philadelphia during the last quarter of the eighteenth century. There are also two straight-sided wing chairs, square-legged with stretchers in the late Chippendale manner. Close by these are placed a couple of adjustable candlestands with the candle arms removed so that they can be used as tables—evidently the original owner's idea. One of them is of the screw type, the other extensible by loosening a thumbscrew. Another eye-catcher is a small, round tripod table with the rod foot common to Pilgrim candlestands.

Situated at the front of the house, where there are few trees, and having three windows, this room can be bright and cheerful, even with all the dark woodwork. However, to encourage the admission of a maximum of daylight despite the two-foot-thick outer walls, the plain curtains are hung on hinged rods so that they can be swung to the side of the reveals.

A somewhat similar, though possibly more masculine, room has the distinction of being dubbed the smoking room, dating as it does from a day when this mystic rite was performed well away from the living quarters and the presence of women. And it is well equipped for that "contemplative" activity, with tiny tripod and quadruped tables, together with an adjustable, ratchet-type twin candlestand, complete with tallow candles, all of them large enough to hold both ashtrays and a glass. Still

The masculine smoking room complete with candles and lanterns.

smaller tripod stands support candle lanterns whose purpose is debatable since they produce little light, suggesting that they doubled as fusees for lighting pipes.

There is very little visual distraction in this room; the walls are paneled all over, and the ceiling is painted blue, while the curtains are a plain beige. Five small hooked rugs add to the old-fashioned air but provide little color relief. The two wing chairs of the Chippendale straight-legged type, one with stretchers, one without, are dated respectively 1810 and 1750–1780, and both are covered in a gaily patterned fabric. There is also a turned ladder-back chair in maple with ball finials that seems to date from the early eighteenth century. On the other hand, a Pennsylvania drop-leaf table with turned legs and center stretcher may go back as far as 1680.

Another fine early piece occupying a corner of the smoking room is a two-drawer, fall-front desk of pine with button feet

An inviting fireplace corner
of the smoking room.

and a deep scroll apron. A melancholy footnote is added by a lifelike though time-darkened model of a parrakeet, enframed by its wire ring perch, hanging in one of the windows—a reminder of the mortality of all living things and the passage of time.

The most recently completed restoration is that of E. I. Du Pont's bedroom at the front of the house on the second floor. This overlooks both the garden entrance court and the tree-lined driveway from the highway known as Buck's Road.

In this recreated room the woodwork is painted a medium blue, the walls finished in a very light tint of that color. The single, fringed draperies, tied back to one side, are of toile in a lovely medallion print suited to the period. This material is also used for the bed coverlet and bolsters as well as the upholstery of an open-arm chair which faces, across the fireplace, a wing chair in blue silk taffeta.

Only four pieces in this room are part of the original furnishings. These consist of a Philadelphia-made Sheraton drop-front secretary-desk of the Federal period, and its attendant black Windsor chair (*circa* 1800), a rush-seated chair left to E. I. Du Pont by his father, and a sleigh bed of mahogany veneer, which dates from 1825–1840. There is, however, now in the room a rush-bottomed rocker which E. I. Du Pont also inherited from his father.

E. I. Du Pont slept here and undoubtedly also wrote his personal letters here.

OPPOSITE: The spacious through-hallway with winding stair.

One of the two rooms that
now form the drawing room.

On these walls are several period pictures, including a pair
of Redute prints of specimen roses which hang over the mantel;
an engraving of a Merino ram imported by E. I. Du Pont in
1800; plus three prints of powder-making equipment from a
French encyclopedia, and a small painting in sepia of Eleutherian Mills done by the Baroness Hyde de Neuville, a friend
of the Du Pont family in France.

When E. I. Du Pont died in 1834, his business office was
transferred from the residence to a new building close by,
later called the Old Stone Office. That building still stands
and contains some of the original furniture that he used.

In 1922, the mills were closed down, and the estate became
the property of Mrs. Francis B. Crowninshield, daughter of
Henry A. Du Pont. It was she who restored the house, part of
which had been used as a club for the workers for almost
thirty years, adding certain modern facilities and some furniture of her own.

The remaining stone structures on the property were preserved, and on some of the surviving foundations classical
ruins were constructed, by Mr. Crowninshield himself, as part
of the mansion's formal gardens. In 1958 the entire estate was
turned over to Eleutherian Mills-Hagley Foundation, by whom
it is maintained as a historic house open to the public.

OPPOSITE: The dining room with its colorful block paper.

ROSEDOWN

Flower of the Felicianas

Rosedown's summer house
in the south garden.

In Louisiana's luxuriant West Feliciana, milk-white, many-pillared Rosedown nestles like some fabulous jewel in a setting of geometric lawns and exotic tree-shaded gardens, its carriage drive arched over by moss-draped live oaks planted 140 years ago. The interiors, too, are much as they were when the house was in its prime, thanks to an extraordinarily thorough restoration—almost a re-creation—carried out within the past decade.

This mansion, whose history reads much like a fairy tale, was erected in 1835 by master builder Wendell Wright for young Daniel Turnbull, a wealthy planter, and Martha Barrow Turnbull, his wife. As a skilled experimental horticulturist, Martha spent the next few years designing and developing the seventeenth-century French gardens that were to make Rosedown House, as it was then called, a showplace of the region. Meanwhile, whenever opportunity offered, the two of them scoured Europe for treasures with which to furnish the mansion and from Italy brought the marble statues that still adorn the garden's avenues and walks.

The residence itself fortunately had always been attractive enough, both in design and elevated location, to serve as a pleasing focal point for all the natural splendor around it. In 1844, however, a curious incident of which Turnbull was able to take advantage resulted in a simple change which gave the house's main façade better balance. What happened was that admirers of statesman Henry Clay, convinced that he would undoubtedly become the country's next President, had a massive Gothic bedroom suite made for him to mark the occasion. When Clay failed to win the election, the furniture was sold to Turnbull who, in turn, was embarrassed to find that the soaring bedposts were too tall for the rooms. He solved this problem by building a one-room wing to receive it, thus, quite incidentally, balancing the library wing at the other end and endowing the structure with greater symmetry.

For the next two decades or so good fortune continued to

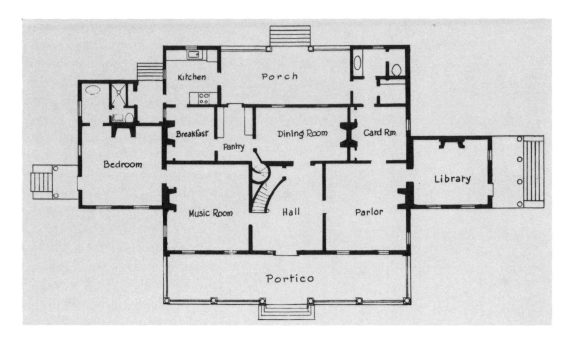

The main-floor plan of the house today with modern washrooms.

The principal façade with its brick-walked parterre.

Fountain view from
the north garden summer house.

The Henry Clay bedroom in
a one-story extension.

smile on the Turnbulls and the three children born to them
since buying the place. Then, with the outbreak of the War
between the States things changed. In the first year of conflict
Daniel Turnbull died, and the family's fortunes began to
wane with the growing poverty around them. Soon there were
practically no funds for the maintenance of either the mansion
or its contents, though Martha continued to occupy the place
until 1897. On her death, the heirs took over and with dif-
ficulty managed to keep the somewhat dilapidated mansion
habitable until 1956 when the estate was acquired by Mr. and
Mrs. Milton R. Underwood from Houston, Texas, who were to
undertake the formidable task of restoring both house and
gardens to their former glory, a project that occupied a small
army of experts until 1964 when the house was opened to the
public as an exquisite relic of America's golden age.

The house, now called, simply, Rosedown, is a particularly
lovely two-story plantation mansion of a style and proportions
one would expect in the early days of the classic revival period—
with certain notable exceptions. For one thing the roof is of
the pitched type, terminating in round-hooded chimneys, and
though the gable ends are clapboarded, the front and rear
walls are covered with flush boarding. The dominating feature
of the principal façade, which adds both dignity and beauty,
is the front porch and the gallery above it, each supported by
six Doric columns connected by exquisite balustrades, and
terminating at the returned ends in pilasters attached to the
house walls. Over the upper columns, the crowning entablature
consists of a shallow architrave decorated with triglyphs,
metopes and guttae, and a nicely molded cornice.

Slightly set back from the main façade are the flat-roofed end
extensions which duplicate, at a somewhat higher level, the
gallery balustrade, square pedestals taking the place of porch
pillars. These extensions with their balustraded roofs naturally
add substantially to the apparent overall length of the structure,
increasing both its air of importance and its architectural
integrity. Six wooden steps lead up from the brick-paved
paths to the house's front porch in which the main entrance is
centered. This is a typical Federal-style doorway with sidelights
spanned by an oblate fanlight. On either side of it are two
pairs of tall sash windows which accentuate their proportions.
This door and window arrangement, together with the trim,
is duplicated on the gallery above.

At each end of the roof peak is a brick chimney with twin
brick cowls. Another chimney, this time with three brick
cowls, is set halfway down each rear gable slope. The two
extensions, likewise, are also provided with their own tall,
hooded chimney, centrally located, plus a pair of windows
matching those of the porch, while a side projection of the
northern extension's roof is supported by two Doric columns
in the manner of a balcony to form a miniature portico. A
notable difference between the two extensions is that on the
southerly one the columns are fluted, while on the later,

246

OPPOSITE: The dining room still boasts its punkah.

The Victorian parlor at teatime.

northern addition, they are not. Other details, such as corner pilasters and cast rainwater heads and spouts, add both to the grace and delicacy of the whole exterior, while the mansion itself is set high enough to dominate the quite formal landscape and plantings surrounding it.

Inside, the house is every bit as interesting as the exterior, thanks to the meticulous restoration in which no pains were spared to salvage every original item regardless of the time and effort involved. In addition, missing items were closely duplicated by pieces of the same period—a practice made possible by the availability of unusually complete records and descriptions of everything bought for the house, together with notes on changes made throughout the years. To facilitate the work done both on the furniture and on the house itself, a temporary workshop was erected nearby in which masters of the various trades rebuilt and refinished the furniture and actually copied all exterior and interior wood- and metalwork, including moldings and trim, that needed replacing.

The actual extent to which restoration and re-creation were called for is well exemplified by the entrance hall with its white-painted classic trim and acanthus-patterned ceiling medallion which provide the architectural accents. Though the curving mahogany stair and the plank floor of cypress are originals, the early-nineteenth-century wallpaper depicting the adventures of Charlemagne needed its border and dado replaced. Luckily a copy made in 1824 by the same artist—Joseph Dufour—was discovered in Paris. The nineteenth-century English needlepoint rug is also a replacement, as is the 1810 Federal, cornucopia-armed settee covered in blue cut velvet, while the bronze chandelier is an English substitute dating from 1840. The marble bust, however, is actually a survival depicting a Turnbull son.

Other originals in this area include a splendid Lannuier console, or pier table, of 1815. This has an inset marble top, the columnar legs and rear pilasters also simulating white marble, the whole piece being ornamented with feet, bases, and capitals of bronze *doré*.

At the foot of the stairs is a doorway leading to the music room whose blue and gold draperies and wallpaper border reflect the major colors used in the hall. One of the most attractive of all the Rosedown rooms, this one relies for its appeal both on the color and pattern of almost everything within it, the major counterpoints being the clear-crystal Waterford chandelier, the crystal candlesticks on the mantel, the sheer net curtains, and the stark white of the walls and ceiling against which backdrop everything else stands out in sharp relief. Particularly effective are the hangings of a serge-like material lined with silk and the black-marble fireplace with its ebony and gilt timepiece over which hangs a large Federal mirror. There is also a lovely rosewood piano, made to order by Chickering, over which hangs an 1821 portrait of Eliza Pirrie by an unknown artist.

Between windows on an adjacent wall is a tall tabernacle mirror resting on an exceptionally ornate console table whose pedestal consists of a gilded grotesque. This piece has the air of a Louis XIV or XV extravaganza poking fun at the formalities around it. However, a major part of the picturesque background is the splendid needlepoint rug in red, beige, and pale blue. This ties together, so unobtrusively, the group of American Empire mahogany chairs with their sweeping lines and lively blue-and-white brocatelle seats. No less important is the little gilt tea table with a Chinese air, such as one might expect to see at the Pavilion in Brighton. Here it nicely balances the brass, folding fire screen while emphasizing the room's focal center established both by the rug pattern and the chandelier above it.

Far less gay yet interestingly colorful is the parlor whose white wallpaper shows little pattern so that the blue-bordered *repoussé* frieze, brass window valances, black-marble mantel, and gilt mirror frames stand out sharply. The red and gold window draperies likewise are given prominence by the white-lace curtains which admit a maximum of daylight and bring out every nuance of color in the tufted red silk velvet of the Mallard walnut sofa and chairs. The greatest mass of variegated color—red, green, blue, beige, and yellow—however, is provided by the round Aubusson carpet dated 1800. The tripod tea table, which wears a white tea cloth, has its jet-black

The library with its early nineteenth-century astral lamp and Argand chandelier.

249

One of the early bedrooms with its massive four-poster.

The Turnbulls' airy chamber in which their children were born.

surfaces almost obliterated by gold-leaf marbling. This table, laid for tea, together with the mantel and a plant-stand vase, displays various porcelain and china pieces, principally Worcester, Rockingham, and Sèvres.

The mirrors here play important parts in extending the view while adding their quota of interest. The overmantel mirror, as wide as the fireplace opening, extends its gold-leaf frame and cresting to the ceiling, while the pier glass on its black-marble base reaches up to the window valances with its urn-and-swag-decorated top. Both fully reflect the light and the sparkling clear glass of the chandelier, which is a duplicate of that in the music room across the hall. Unluckily they cannot reflect the most interesting piece in the room—a needlepoint fire screen worked by Martha Washington and enframed in a mahogany tripod stand, pierced and carved, and incorporating the most intricate of curves.

To the rear of the entrance hall is Rosedown's dining room, whose most nostalgic feature is the ceiling fan, or punkah, over the dining table. This punkah, incidentally, had to have its original needlework covering replaced, but the room still has its Turnbull furniture, the saber-legged chairs being a Late Empire style with curled backs and carved splats, possibly by Phyfe. The table, with its elaborately carved legs and its casters, is a drop-leaf extension type of the same period. The side table, on the other hand, is a marble-topped pier table of mahogany, ornamented in bronze *doré* but much less fanciful than the one in the hall. In this room, too, is an original black-marble fireplace with Ionic columns, supporting a huge, gold-framed mirror. Over this is an elaborate paper border in red, blue, and gold, which contrasts sharply with the pinkish-beige wallpaper and white ceiling. The windows in this room are draped to one side, in mauve silk with tassels, over glass curtains of white lace.

Since there is no chandelier over the dining table (the punkah would prohibit that), there are triple candlesticks in ormolu on the mantel, French bronze and crystal wall sconces with flat, pear-shaped amethyst drops, and on the side table, a bronze and crystal twin Argand lamp. On the dining table are the usual candlesticks, this time of crystal.

Besides family portraits by Sully and a pierced-brass fender, decorative features include displays of Flight, Barr and Barr dessert plates on half of a four-stage dumbwaiter standing flat against the wall, with others on the mantelpiece. The largest mass of color and pattern is provided by the room-sized Aubusson rug in gold, red, and yellow on a pale-beige ground, a scheme guaranteed to tie the whole room together.

The use to which the small room, now furnished for breakfasts and informal meals, was put is now doubtful, but it has been endowed with a charm of its own. In it harmony has been achieved by copying the original wallpaper, which is green and gold on a white ground, painting the trim white, and the ceiling off-white, using white silk curtains from India, with

OPPOSITE: A corner of Rosedown's music room.

a fine gold thread running through them, under pelmets of green silk taffeta with tassels. Other touches of green are supplied by a pair of enameled pole screens with gilt decoration by painting the fireplace facing, and by color bands on a pair of Old French urns which, together with a gilt clock, form the mantel's garniture.

The tile-patterned rug made by Aubusson for this room also has touches of green so that the green-painted, flower-adorned chairs from the Brighton Pavilion seem perfectly at home. The table is an Empire-style folding mahogany piece. Above it, with its anthemion decoration, is a gilded wood and crystal chandelier whose image is almost completely reflected in the gilt Sheraton mirror over the mantel, which incorporates an *églomisé* panel in gold and black.

One of the most inviting rooms in the house is the library with its highly personal desks, books, and writing materials, its mercury barometer (to be tapped ever so lightly every morning), and the comfortable seated pieces. The desk is fascinating; its large, neat, and businesslike foursquare top, with four ample drawers, contrasts vividly with its carved and tortuous legs and looping stretchers tied into a decorative central pedestal. On it stands an astral lamp, but the room's lighting was left to a four-branch Argand chandelier. The three French walnut bookcases made for this room add to the architectural character of the interior. More interesting is the fact that they contain Turnbull's books.

Facing the window is a plain marble fireplace with a nineteenth-century portrait of Edward Livingston by Vanderlyn above it, while on the mantel stands a brass clock with gilt decoration and a pair of short candelabra with tall chimneys. Below is a handsome fender of pierced brass, with tall brass andirons behind it, while on either side is arranged an upholstered seat with a rounded, sloping back, the two when put together forming a settee. The chairs are saber-legged Empire pieces, but the pair of globes (celestial and terrestrial) with tripod bases are somewhat earlier.

With so much walnut in this room, and the other dark woods, plus the heavy draperies and deep pelmets of crimson and gold damask, it was obviously necessary to dispel the gloom by making the ceiling and walls white (though the latter are papered in wide stripes with a faint floral pattern) and using an Aubusson rug having a white and light-beige ground with a scattered pattern in comparatively quiet hues.

The most important bedroom is, of course, that furnished with the Gothic suite made for Henry Clay. Since the room was designed to house these pieces, the architectural detail, such as the frieze and cornice, the door and window trim, and the marble fireplace, were all planned to form a suitable setting without emphasizing the Gothic detail which was overpowering enough. The doorframes, therefore, were monumental but plain—the upper door panels two-thirds the height of the door; the heavy lintel capped but plain. The frieze and cornice

The breakfast room with its tile-pattern Aubusson rug.

OPPOSITE: The classic entrance hall and high, steep stair.

253

The nursery replete with old-time playthings.

The Gothic bedroom furnished for a possible President.

moldings are deep, beginning with large dentils, then tongue and dart, followed by a deep cavetto with a beaded border, and topped by a row of miniature petals. The wallpaper carries a series of riband and garland borders in gold forming vertical panels on a white ground, faintly stippled.

On this floor is a needlepoint rug with a pattern of large, round-cornered squares separated by red and gold irregular outlines in which the corner spaces are ornamented with avellan crosses in gold. These rounded squares carry floral designs in gold, black, and red—the red matching that of the chair and ottoman fabric which carries a pattern of gold wreaths.

In sharp but interesting contrast with this very definite red, the bed canopy is of a medium-green figured material, outlined in red, with a central gold tassel guaranteed to hypnotize the reluctant sleeper. The bedhead hanging and the quilted coverlet, however, are of a deep golden color, the coverlet having a central checkerboard pattern of small, red squares, the whole bordered by diagonals in the same tone. Strangely enough, the old gold and green harmonize well with the mahogany, all tied together by the larger areas and smaller accents in white.

Since the great bed stands in a corner between two windows, each pair of translucent white curtains with red bobble fringes is hung from a carved wood replica of the bed's canopy molding. This effectively unifies room and furniture.

In addition to the massive 13-foot 6-inch-high bed there is an armoire, an extremely heavy cheval mirror, and a carved washstand besides the arch-backed chairs. Even the tiny night table has its pillars and drops; only the fireplace is plain, being content with a frame of wide ogees to tie the mantel to its inner facing. Above this is a definitely nonecclesiastical mirror in an elaborate gilt frame, double-bowed at the top to accommodate an intricate and extremely ornate cresting with garlands looped across the curving corners.

Flanking this on the mantel is a pair of circular, fanlike candle shades on tall, metal stands, used in olden times to shield the face from candlelight. The rest of the garniture consists of a clock in the form of a gilt metal Gothic church and a pair of brass-footed candelabra with etched globes. Other crystal-hung candle branches are fastened to the 6-foot-tall ogival dresser mirror, and an astral lamp with a gilt base, porcelain shaft, and crystal-hung shade perches jauntily on a small bedside stand. There is dignity here, and a certain chaste atmosphere, but also solid comfort.

It is interesting to compare with the Henry Clay room the airy chamber above the parlor which the Turnbulls selected for their own, and in which their children were born. The Sheraton "swan" cradle, *circa* 1820, however was bought in England more recently. On the other hand, the great rosewood bed, the ponderous dressing table, and the washstand were made to order by Prudent Mallard, the noted cabinetmaker of New Orleans.

The Henry Clay Room's fireplace
and elaborate mantel mirror.

The rest of Rosedown's rooms are of almost equal interest as exemplifying interiors likely to be found in an early-nine-teenth-century plantation house in which cost was a matter of little importance. The thing to be considered here, however, is the remarkable extent to which restoration and preservation for the future have been carried, the result being not merely a museum but a mansion in which every item is as functional as it was when first acquired. Furthermore, its maintenance in good condition is assured by the fact that the house itself is fully air conditioned—climate being a worse enemy than time. Every room has its sprinklers, each roof its lightning rods, and the whole of the mansion a complete alarm system, each installed in such a manner that few would notice its exis-tence—or even that all lighting fixtures have been electrified, with modern light bulbs that simulate the flame of a candle or a lamp. These utilize an unfailing source of current so that the fascinating interiors that resounded for so long with music and the voices of children may be made visible on the darkest days—or nights—for centuries to come.

Mobile, Alabama

PALMETTO HALL

Begun 1846 – Transformed in 1962

Palmetto Hall with its two
modern end extensions.

What was to become one of the South's finest classic revival mansions was erected in Mobile in 1846, though it did not achieve that enviable distinction until 115 years later when it was not only transformed, inside and out, into the loveliest of residences, but had its floor space almost doubled in order to serve both as a comfortable home and as a repository of some of America's finest examples of native art and craftsmanship.

Originally, the house, Palmetto Hall, consisted of the present central block, two stories high and almost square in plan, with a wide exterior staircase to the second floor where the family's principal living quarters were then located. At that time the upper floor was divided into four principal rooms, a living room and dining room on one side of a central hallway and two bedrooms on the other. Downstairs the room arrangement was somewhat similar, though adapted to different purposes, and the decorative details far less elegant.

In the hurricane of 1906, the outside stair was demolished and other external damage inflicted, but complete rehabilitation was not attempted until the estate came into the possession of Mr. and Mrs. Jay P. Altmayer in 1959. At that time a thorough remodeling was undertaken, with the help of architect Myrlin McCullar of New Orleans, preserving the original fabric insofar as possible while replacing the outside staircase with an inside one, adding a single-story wing at each end to increase the living space, and modifying the interiors to suit, an undertaking that occupied almost 3 years. At the same time some of the decorative details were altered to conform more fully with the classical features inherent in the original design. The overall result was a unified structure, perfect alike in details and proportion, without the slightest loss of its antiquarian flavor from whatever angle it is viewed.

In the manner of so many Southern plantation houses, the principal façade is sheltered by a deep portico, beneath a second-floor balcony which duplicates its proportions. While the balcony is supported by stout Tuscan columns of plastered

The mansion from the rear lawn.

brick, complete with necking and square base, the roof entablature rests on beautifully proportioned fluted wooden columns of the Doric order directly above them. The frieze, which is adorned with triglyphs, metopes, and guttae, under a deep, molded cornice, is carried right around the roof of the central block of the house, adding tremendously to its classic air. A feeling of solidity is added by the smooth wall of the main façade which is entirely of stuccoed brick.

On the ground floor much of this wall area is exposed, because of the fact that the openings, or voids, are limited to a single French window on either side of the double doorway, with everything painted white to match the wall except for two pairs of floor-length blinds. The main entrance door, as a matter of fact, is solid-paneled, set between two slender pillars inside the architrave, the overdoor and frieze being decorated with paterae. Low stuccoed walls tie the façade to each end column, adding an air of compactness and solidity to the porticoed area where it needs it in view of the sturdy aspect of the low, setback wings.

Upstairs, a different note is struck with the two pairs of windows flanking a paneled doorway whose semicircular fanlight spans a pair of sidelights, each set between attached, fluted pillars. Seen through a framework of moss-draped live oaks, the whole central unit sparkles in the sunlight like an exquisite ivory carving, meticulously detailed and faultless in

Under the dining room's Federal girandole are chairs inlaid with silver decorated with an eagle and sixteen stars.

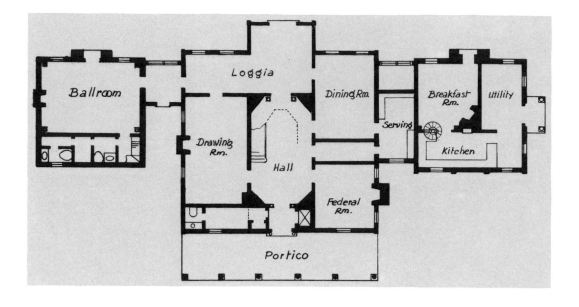

Floor plan of the original mansion and its 1962 additions.

The Federal Room's white marble fireplace with supporting terms.

design, the half-seen wings dappled in shadows, promising more wonders to come.

Actually, not until one moves around to the rear of the house can the unity of the wings and the central block be fully appreciated. Here, where for half the day sunshine floods the wide lawn, one faces an incredibly long façade centered on a slightly raised, flagged patio the full width of the central block which was the original house.

In the low, balanced wings and the hyphens connecting them to the central block, it is possible to recognize a touch of the Palladian, emphasized by the low-pitched, pyramidal roofs crowned with a pedestal-like stone structure that serves as a chimney. In the center of its rear façade, each of these flankers has its round-topped entrance doorway in the middle of a pedimented pavilion, with a single window at each side set into the main wall. Thanks to the beautiful proportions of each element of the structure, it was quite unnecessary to apply ornament in any way, a natural exception being the œil-de-bœuf window enclosed in a laurel wreath within the tympanum of each pediment, a treatment duplicated in the louvered screen of the hyphens.

The rear façade of the central block has been transformed by an addition of a deep pavilion which not only gives it more character but also provides a little more space in the interior. The pavilion is equipped with a wide, wrought-iron balcony on brackets at second-floor level where a pair of tall French doors, topped by a semicircular fanlight, opens onto it. Below this is the main-floor garden entrance whose twin folding doors, almost of ceiling height, incorporate four full-length glass panels, duplicating the design of the glazed balcony doors. Decorative features on the flagged terrace consist of stone urns on pedestals flanking the great doorway and almost life-sized sculptured figures where the patio steps lead down to the lawn.

Immediately inside the front entrance to the mansion stand

The unusually shaped breakfast room with its pewter-inlaid mantel.

a pair of ceiling-high Corinthian columns with bases and capitals of cast bronze. From this point the walls open out at forty-five degrees to the full width of the entrance hall, where a cantilevered stairway sweeps up to the second-floor landing via another forty-five-degree turn. With only a narrow landing to one side of the stairwell, the whole of the 40-foot-high, deeply coved ceiling is visible from the lower hall—an elongated octagon bordered by a finely molded cornice. From the center of this hangs, at the lower ceiling level, an exquisite Waterford chandelier.

At the rear end of the hall another pair of Corinthian columns flanks the doorway to the loggia whose quadruple glass door immediately provides a view ahead of the rear terrace and the tree-dotted lawn beyond. Against the white walls and trim of the entrance hall, the dark pine of the floor, and the beige ground of a huge Savonnerie rug, each picture and each piece of furniture, spaced well apart, is properly accented. These pieces include George II armchairs in mahogany; a pair of tapestry-covered love seats of the same period; and an Adam marquetry side table, each with an important painting above it from the Altmayer Collection of the works of American artists. Displayed here are such important portraits as Rembrandt Peale's *Washington* and D'Avignon's *Winfield Scott*, while at the top of the stairs Henry Clay is pictured in *The Great Debate* by John Nagle. On the opposite wall angle, another portrayal of Scott is shown, this time by Thomas Sully. One of the hyphens leading off the hall accommodates a group of approximately forty Southern-life paintings by William Aiken Walker.

Stepping into the loggia involves also entering a cross hall leading from the dining room to the ballroom. Here the whole floor is of white-marble inset with a narrow border in mottled black. The walls are punctuated with pilasters decorated in stucco, with a carved frieze painted white to match the ceiling, in pleasing contrast with the beige walls and *café-au-lait*-colored draperies. The rugs are elaborately patterned Kirmans in a medley of colors, the whole illuminated by Louis XVI chandeliers, which can be augmented on occasion by Regency-style triple wall sconces, carved and gilded around cameo centers.

Among the loveliest furniture pieces are two George II eagle consoles and a sumptuous Regency marble-topped commode, with mirror-paneled doors, on which is displayed a Remington bronze, *The Bronco Buster*, and over this is hung an 1832 portrait of General Drumgolle by Sully. In the floor's center is a Regency octagonal platform table on dolphin supports with winged lion-paw feet in gilt bronze. The largest piece, however, is a quite modest English chest-on-chest, *circa* 1740, with a bust of Washington by Houdon on its top. Here also are paintings by G. P. A. Healy and Albert Bierstadt.

The living room differs from most of the other interiors by being paneled. All the overdoors and the overmantel, which were made in London around 1760, are resplendent with

The intriguing Beau Brummel, or Baltimore chest.

OPPOSITE: The entrance hall with its Waterford chandelier.

The Palmetto dining room.

heavy triangular pediments having a large cartouche in the center and a deep pillow block below it. Between the pillow block and the mantelshelf is a wide, egg-and-dart molded frame, with ears at the top inset with paterae and scroll brackets at the base. Inside this frame is mounted a gold-framed Sully portrait of Esther Cox Binney, dated 1832. Quite interesting also is the 1780 wooden mantel, salvaged from a Long Island house, whose central panel is inlaid with a Grecian figure in pewter, the ends supported by Ionic pilasters enclosing a beige-marble facing. With colorful Derby vases as mantel garniture, the whole projecting chimney breast, from broken cornice to hearth, constitutes a stunning ensemble.

The room's principal lighting fixture is a fine Bristol chandelier in crystal, local illumination being provided by assorted vase-based table lamps. Apart from two or three modern upholstered pieces, the furniture is wholly of the George II period (1727–1760), set off by a signed Kirman rug and draperies in beige silk damask. Possibly the most interesting piece is the tall bookcase in faded mahogany which was made by William Vile (died 1767), which bears the crest of the Sargent family, though the Regency consoles alongside the double doors display an equally rare pair of porcelain fruit coolers by Flight, Barr and Barr, who owned the Worcester pottery from 1813 to 1840.

The ballroom is also of particular interest, since it is unusually well appointed for an interior dedicated primarily to such a purpose. Attached Ionic columns at the four corners, carved overdoors, and a white-marble mantel with supporting caryatids add to the formality, as do the draperies of brown cotton sateen of a medallion pattern in yellow and blue, with their heavy swags and square-hung cascades, as well as the room-size Oriental rug over a floor of white-marble tile.

Within this setting a number of fine furniture pieces are arranged around the walls, some of them small but good, such as the Charles X bronze mantel clock with its finely molded Indian maid with bow and arrow, in bronze, and the pair of Louis XV porcelain urns atop a huge Regency sideboard table with a scagliola top. The only piece centered on the rug is a tilting Regency sherry table with a brass-inlaid circular top, dating from somewhere between 1805 and 1810. The vase-back armchairs are Portuguese adaptations of Queen Anne pieces, but the walnut twin-back settee with eagle-head arms appears to be a nineteenth-century copy of a George II piece. Among the more important paintings is a portrait of General U. S. Grant by Nicolai Marschall, painted in 1862, that of a lieutcnant in the War of 1812, attributed to Charles Polk, and a magnificent portrait of Zachary Taylor by James Henry Beard.

A William Vile bookcase
in the living room.

The sword collection
in the Federal Room.

The breakfast room as seen from the pantry.

The living room and its pewter-inset mantel.

One of the most interesting rooms, which gets its name from the period in which it is furnished, is the Federal Room. This has draperies in gold silk damask against walls of a warm yellow, a color repeated in the anthemion decoration of the wide frieze, which, like the ceiling, is dead-white. Besides creating a sunny atmosphere, this tint also emphasizes the contours of the white-marble mantelpiece with its supporting terms and elegantly detailed frieze which enclose a bronze fire grate and its molded bronze frame. The *garniture de cheminée* is just as exciting with its two tiny Wedgwood urns ornamented with cameolike medallions displaying busts of William Pitt, George III, Benjamin Franklin, and Alexander Hamilton. Above these is a portrait of George Washington by Jane Stuart.

Undoubtedly the most fascinating piece of furniture in this room is the Baltimore desk, or Beau Brummel. This combines a pigeonholed writing desk with drawers, a makeup drawer with a hinged looking glass, and a glass-fronted display cabinet with a clock centered in its pediment. It is made of mahogany with bronze-mounted quarter pillars, beautifully banded especially around the doors and pediment, the latter furnished with eight pineapple finials. Through the round-topped central door can be seen a collection of silverware, the smaller side-door panels being decorated in *églomisé*.

All of the chairs are Biedermeier pieces in old-gold cut-velvet upholstery, having veneered arms terminating in rams' heads of bronze. There is also an inlaid round table, with bronze phoenixes as pedestal supports from the same atelier—the whole group held together by an Aubusson rug in a pale beige, soft tans, red, and hyacinth blue.

The loggia and its period French chandeliers.

The ballroom in its sitting-room dress.

The master bedroom.

A major feature of the Federal Room, however, is a glittering collection of presentation swords in silver and gold, set with precious stones, mounted in a case that occupies one wall. Each of these swords was owned by a notable historical figure, ranging from General Grant, Winfield Scott, and Zachary Taylor to Admiral Farragut and Schuyler Hamilton. Also reminiscent of great historical events are the room's paintings.

Another interior that is decidedly different is the oddly shaped breakfast room, which is paneled in cypress and has a pewter-inlaid mantelpiece which duplicates that of the living room, though in this case it is faced with antique tiles. Over the mantel is a painting of a United States troop transport in the Mexican War by Edward Arnold of New Orleans, the facing wall being dominated by an 8-foot eagle in gilded cypress, once a feature in the main saloon of a Mississippi riverboat.

This room is floored with bricks which were a part of the original 1845 structure, and in keeping with its early American air, the interior is furnished with Windsor chairs, four of which are grouped around a circular table on an oval hooked rug. In a corner where the room opens into the butler's pantry is a circular iron stair to the second floor, an intriguing addition to an altogether fascinating mansion that has grown so tremendously both in beauty and utility over the past 120 years.

Williamsburg, Virginia

THE GOVERNOR'S PALACE

Colonial Center of Sovereign Power

In the perfection of its simplicity, the early Georgian mansion built at Williamsburg in 1705–1722 to serve as the governor's palace proved an architectural triumph despite its moderate size. This is due to its elegant proportions and the felicity of its architectural detail, which eliminated the need for any decorative features save one—its crowning glory, the gleaming white cupola atop its tall roof.

With neither pediment nor pavilion, with no wide portico to emphasize its main entrance or the long, curving colonnades that tie a Palladian villa to its flankers, it achieved a dignified and even impressive façade—regardless of a certain tight-lipped austerity suggested by the plain and narrow doorway—quite in keeping with its importance as residence of the king's representative in the then vast Virginia colony which extended west to the Mississippi and north to the Great Lakes.

Set within ten acres of gardens, lawns, and terraces, the original mansion was a mere 60 feet long, a little less from front to rear, and two-and-a-half stories high over a raised basement that set it six easy steps above the 100-foot-wide forecourt onto which it faced. This graveled area was flanked by a pair of single-story brick buildings whose gable ends faced the street, the one to the west serving to house the palace guards, while the other accommodated the governor's business offices. Along the front boundary of the forecourt was built a stout brick wall, its curving ends sweeping inward to connect the two flankers. At its center this wall was interrupted by a pair of massive brick pillars surmounted by the royal lion and unicorn, carved in stone. Between them swung double iron gates whose crests bore the royal cipher in shining gold. This entrance was narrow enough to admit nothing wider than two people side by side to the flagged path leading to the palace steps, a circumstance that would necessitate even the governor alighting from his carriage to walk to the front door regardless of the weather.

Such was the stately eighteenth-century mansion designed for

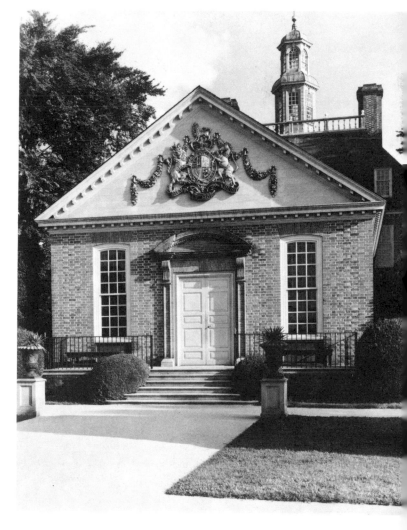

The supper room behind
the palace.

The palace with its flanking dependencies.

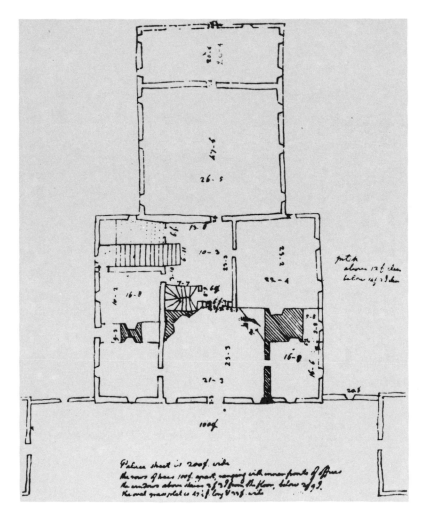

Jefferson's sketch plan of the Governor's Palace.

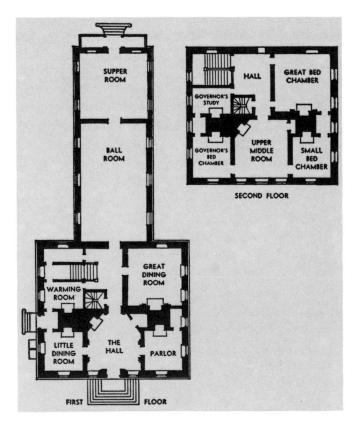

A plan of the palace restored.

268

OPPOSITE: The ballroom with its coved ceiling and tall windows.

Governor Alexander Spotswood—himself no mean architect—as his official residence and erected by master builder Henry Cary. Not until 1752 were any important changes made to the main building. Then, in 1753, a narrow, single-story extension, 75 feet long, was added to the rear of the palace, accommodating both a ballroom and a supper room for formal affairs. Unluckily for posterity the central block of the palace was destroyed by fire in 1781, the smaller, flanking buildings surviving until Civil War days when they, too, were razed.

Thanks to several years of research by experts in the nineteen twenties, accurate reconstruction was made possible, utilizing the original foundations as checked against a dimensioned floor plan sketched by Thomas Jefferson while governor in 1799. Evidence of the external appearance of the palace was provided by a copper engraving found in Oxford's Bodleian Library, plus many public and private records that survived, covering not only the structural details but also listing the furnishings and interior finishes at various periods in its history.

Erection of the buildings was begun in 1930 and completed four years later. As rebuilt it constitutes an exact copy of the original palace, actually incorporating the old basement and wine cellar which were found intact. The present structure therefore consists of a T-shaped building composed of the almost square unit housing the governor's quarters and the later addition which terminated in ornamental gardens at the rear.

One of the details responsible for the handsome appearance of the principal façade is the fact that, like the rest of the exterior, its bricks are laid in Flemish bond (*i.e.,* alternate headers and stretchers) which give the surface an unusually interesting texture. This wall is divided horizontally by a brick stringcourse which emphasizes its width. Below this is the slightly elevated entrance with its graduated stone steps; the door itself, being narrow, with its trim simple, is quite inconspicuous even though painted a light gray. This humble portal is flanked by two pairs of ceiling-high windows, spaced well in from the building corners, which are free of quoins or any other decorative touches.

At second-floor level the doorway is sheltered by a rectangular, iron-railed balcony (in lieu of a pediment), onto which it is possible to step through a wide, central gib window whose lower section is enclosed by solid doors to windowsill height. Above the upper windows is a projecting modillioned cornice, marking the lower edge of the roof, which climbs steeply to a balustraded deck. From the center of this deck soars the octagonal cupola topped by a domed lantern whose gilded windvane seems to touch the sky.

While the front slope of the slated roof carries five dormers, the other three sides sport only four apiece, each jauntily wearing its hipped slate roof like a hat. Springing out of the roof close behind the deck ends are two enormous pilastered chimneys in brick, as solidly erect as sentries at attention. Far below is the roof of the ballroom extension which terminates in

The parlor at teatime.

OPPOSITE: The formal supper room—almost wholly Chinese in influence.

A glimpse of the formal dining room.

The royal coat of arms in the entrance hallway.

a huge pedimented gable adorned with the British royal coat of arms in high relief. This gable oversails the end wall, whose wide double door is flanked by tall windows looking out onto an iron-railed stone terrace from which a flight of steps leads down to the garden walk. Viewed from the rear garden gate, this whole composition, with the main building as backdrop, appears only slightly less formal and impressive than the front. Viewed from the palace green, more elements combine to form the overall picture of the place as it must have been two centuries ago—the principal façade, the gables of the flankers, the outer wall, and the decorative gates facing onto a wide carriage ring intersected by foot paths, the grassy areas all around them protected by substantial painted posts and rails, some of the posts linked by heavy iron chains. And nearby, close to the gates stands a single, octagonal, stone mounting block, patiently awaiting a horse and its rider who never appear.

The front entrance to the palace is by a pair of four-paneled painted doors over which is a rectangular ten-light transom, all enframed in the simplest of casings—a sober introduction to the splendors within. In the six-sided entrance hall it immediately becomes obvious that this is no simple burgher's home; the eye at once leaps to the rack of bayoneted muskets, the starlike display of naked swords, and the huge coat of arms, carved from oak in high relief, adorning the dark walls above a fireplace of white marble, whose center panel is carved in a floral design. The floor, too, is of white-marble inset with decorative groupings of black-marble squares inside a border of similar pieces.

In strong contrast with the floor are the walls of black walnut with their raised panels, heavy dado, and tall, fluted Ionic pilasters upholding a heavy molded frieze and modillioned cornice. Flanking the door that gives access to the family dining room stands a pair of late-seventeenth-century tall-back English chairs of the Restoration period, heavily carved in the Baroque manner, which add to the room's impressive air.

To the left of the entrance hall is the "little" or "family" dining room, where informal meals were served and intimate friends entertained. These beautifully paneled walls are painted a rich gray-green, as is the elaborately carved fireplace mantel, whose facing of pink marble is inset with a white keystone and crossetted within a narrow decorative molding. Over it hangs an imposing portrait of Governor Spotswood, attributed to Charles Bridges.

The furniture here is mostly English, including the splendid sterling-silver chandelier, made between 1691 and 1697 by Daniel Garnier. The silver candlesticks on their stands of carved wood and silvered gesso with japanned tops are rare survivals of the Charles II period (1660–1685). A somewhat similar pair of candlesticks adorns the dining table. Notable also are the Chippendale-style oval, drop-leaf table, displaying a porcelain Chinese Export bowl, and the six Queen Anne chairs with their original needlework seats, four of which surround it. In the corner between the doorway and the mantel is an English tea table in the Queen Anne style, while the mantelpiece is garnished with pieces of Whieldon agateware.

At the opposite side of the hall is the parlor which matches the little dining room in size and arrangement (in mirror-image fashion). This is furnished wholly in eighteenth-century antiques, set off by pale-cream walls and golden damask draperies—tints that are repeated in the chair and settee coverings—which combine to enhance the colorful pattern of the English needlework rug. On each side of the curtains is a walnut-framed looking glass enriched with gilt gesso decoration.

From the ceiling's center depends a small English-glass chandelier, whose light can be augmented as required by a pair of crystal-hung sconces with finely etched globes astride the mantel. Set in the overmantel panel above them is a large portrait of Evelyn Byrd of Westover whose father, it is said, once jestingly dubbed her an antique when she remained unmarried at twenty-one! This fireplace is of white marble, inset with scagliola panels, the center one, which was found on the site, decorated with a marble medallion depicting a deer in a forest. On the mantel shelf is displayed a collection of Chelsea porcelain birds. The furniture includes a Chinese Chippendale settee and two

The royal cypher—a feature of the ballroom.

Detail of wallpaper and carving in the supper room.

armchairs, once owned by John Wentworth, the last royal governor of New Hampshire. These lend a touch of the Orient to a room predominantly English with its piecrust-topped tripod tea table, a little urn stand with a fretted gallery, and a large Chippendale folding game table which makes an interesting console.

Beyond the hall there are still two other rooms, the "great" dining room, used on state occasions, and a tiny warming room where food from the outdoor kitchen was kept heated for serving. Between them is the main passageway housing the principal stairway and a smaller, winding one for the servants.

The large dining room, where important guests were entertained by the governor and his lady, is not only the ultimate in formality but rich in its architectural details and somber in its coloring. As in the other rooms, the walls are paneled and the cornice modillioned, but here these are reinforced by fluted Corinthian pilasters on pedestals placed on either side

The little dining room.

The governor's office.
Note the dispatch box.

of the fireplace, their projection being carried to the top of the cornice. These are repeated at the side of the chimney breast, accentuating the three-dimensional effect.

The tall mantel itself is chaste in design, being of white marble entirely without decoration except for plain moldings around the single horizontal panel and the eared facing. On the shallow mantelshelf are eleven Chelsea porcelain figurines, in accordance with Governor Botetourt's inventory of 1770, and in the panel above them hangs a portrait of King James I, attributed to the Italian painter Fredrigo Zuccaro, *circa* 1543–1609.

The elaborate pelmets and rich draperies in this room are original seventeenth-century items of cut velvet in dark green and ivory which do much to establish its air of affluence. The antique Axminster rug in black, red, and yellow forms an excellent background for the mahogany furniture, the gilded Rococo mirror and marble-topped gilt console, the needlework-covered side chairs in three different designs, and the exquisite silver pieces, which include English wall sconces of the William III period and a double-pagoda epergne of 1759 made in England by Thomas Pitts, now gracing the long multilegged mahogany table.

The largest room in the palace is the ballroom, though even this is far from monumental, being but 47 feet 6 inches in length. It is, however, one of the most beautiful from both architectural and decorative standpoints, its great height being scaled down by a coved ceiling and a cornice offset to frame the deeply recessed windows. Its acoustic qualities are said to be superb.

The governor's study.

The northeast bedroom with its Elizabethan bed.

The upper middle room.

Both the window recesses and heads below the cornices are paneled, as is the entire dado, and at the base of each window is a recessed seat with a red cushion to match the color of the cornice-high draperies of silk damask. At each end of the room is a wide, double-leaf door of mahogany, whose fielded panels are edged with egg-and-dart molding. Each of these doors is set in an eared surround flanked by paneled pilasters which terminate in decorative brackets. These vertical modillions support a huge scroll pediment whose tympanum is enlivened with an elaborate cartouche, one of which displays the royal cipher of the Georges, the other the British royal coat of arms.

In this room the walls and ceiling are of a pale blue, all the woodwork white, providing a background that sets off the fabrics and furniture to perfection. On one side of the doorway to the hall is a large portrait of George III; on the other side a likeness of his consort, Queen Charlotte, both painted by Allan Ramsey. At the other end of the ballroom the paintings are of Charles II and *his* Charlotte. Under the former stand side tables of Chippendale's Chinese-period design. Other features of equal note are the three chandeliers of English crystal with bell-and-cup stems. Around the walls are spaced Queen Anne chairs and stools in walnut with seats of red damask. Similar chairs for the use of the musicians are grouped around a double-keyboard harpsichord made by Jacob Kirckman of London in 1762.

Beyond the ballroom is the formal supper room, whose chief glory is a Chinese wallpaper of a type used in England and America in the eighteenth century. Its basic colors are gray on pale blue, with black-winged birds as accents. This Chinese influence is carried over into the carved woodwork and cornice, the door pediments, and a Chippendale-style china shelf in black lacquer. The chairs are a mixture of Chippendale and Queen Anne styles, most of them with cabriole legs and elaborately carved splats, though there are some very interesting variations with gilt gesso decorations that were made in London.

Important architectural accents are provided by the unusually decorative dentiled frieze with its punched, lacelike border and lightly carved cornice, by the window surrounds whose crossettes are bordered with decorative moldings, and by the elaborately carved scrollwork above the dado. Entirely inside this enframement hang triple-swagged draperies of silver lampas.

No review of these interiors would be complete without a brief glance at the family's private sitting room on the second floor, which is also known as the Upper Middle Room. This is one of the most luxurious apartments in the palace, its walls covered with tooled Spanish leather of the 1680's, providing both exotic texture and pattern as a backdrop for products of early craftsmanship, such as gilt-gesso girandoles, tall Queen Anne pier glasses over gilded, marble-topped consoles, and English cut-velvet draperies in gold and brown. Here, too, is a rosewood and mahogany bookcase—possibly by Giles Grendey—and an equally splendid tall case clock, *circa* 1699, by Thomas Tompion.

All the interior woodwork, from window trim to paneled dado, is in walnut, the chairs of japanned beech. On an oval table in the center of the floor is a fascinatingly complicated seventeenth-century Jamaican casket with tortoiseshell and mother-of-pearl inlay—a gift from Britain's Queen Elizabeth II and a reminder, if any were needed, of the palace's ties with the English throne in Colonial days.

The governor's bedroom.

The entrance hall.

Florissant, Missouri
VOUZIERS

French Elegance in the Great Midwest

The main entrance,
guarded by Gardet's bronze dogs.

One of the most stately of all styles of domestic architecture is that of the seventeenth-century château which, while avoiding the fussy details of the Baroque, still is crowned with that fascinating combination of steep slate roofs, *œil-de-bœuf* windows, and quadrant-topped dormers—reminiscent of historic places such as Malmaison and Denmark's Amalienborg—details that continue to emphasize not only the importance of the roof as an architectural entity, but also the distinctiveness of each story beneath it.

Vouziers, on its high bluff overlooking the confluence of the Missouri and Mississippi rivers, is not only such a mansion, but it has, in a literal sense, much more to it than meets the eye. Not only is one of its five floors underground—with a cellar beneath that—but also, hidden under its cascading lawns, which sweep down in giant steps from one level to another, is a huge ballroom, together with its fresco-decorated access tunnel and carport. The exposed façade with its stone balustrade, reached by semicircular steps at either side, looks out upon a massive stone fountain with a spring-fed swimming pool far below.

This "burying" of the auxiliary buildings obviously eliminates any undesirable interference with the view from the low-set mansion or its terraces, despite an assortment of aboveground structures, such as colonnades and *pavillons*, and their connecting, treelined *allées* which are all carefully disposed between the mansion and the Missouri river with this factor in mind. Beyond this complex, the borders of the estate extend from the river escarpment inland to extensive wooded areas beyond which, on a clear day, the towers of St. Louis, fifteen miles away, can be descried.

There are two aspects of Vouziers which some may find difficult to reconcile. One of these concerns the aboveground part of the mansion—the *rez-de-chaussée,* as it is called in France—which, inside and out, re-creates the atmosphere of the days of the Bourbons, while providing every conceivable

OPPOSITE TOP: Nature plays a decisive part in this view of the principal façade.
OPPOSITE BOTTOM: The gallery—a broad hall with an intricately patterned parquet floor.

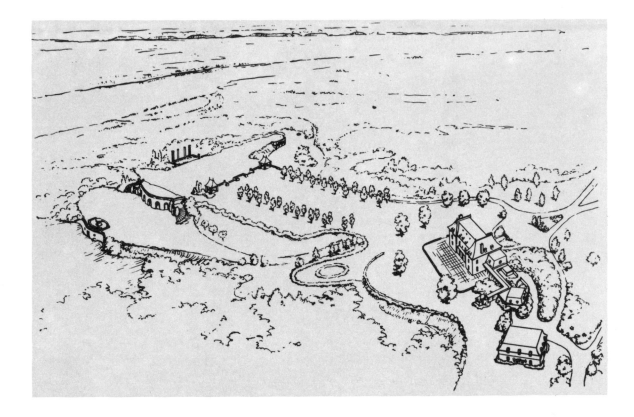

The estate faces the
Missouri River.

modern facility for the enjoyment of existence on the highest
cultural and social levels. The other is the *sous-sol*, the floor
below ground level, where a quite different living space, even
more fascinating because of its remarkable interiors and
exquisite though unusual furnishings, is reserved for large
social functions and the more intimate occasions.

This, then, was the great house that was designed and built
for Mr. and Mrs. Joseph Desloge in 1925. Remodeling changes
and additions were made by Harry I. Helmuth of St. Louis who
was responsible for much of the fabulous carved stonework of
the underground rooms and the various supplementary
buildings, as well as the decorative features at the foot of the
grande allée leading from the house to the theater, the *pavillons*,
and the stone balcony enclosing the outdoor dancing area and
the terraces above the buried ballroom.

The exterior walls of the house and the interior walls of the
vestibule and the *sous-sol* are of Indiana limestone which has
retained its pristine look to this day. The house's principal
façade is a particular delight, terminating as it does in projecting pavilions with slender, hipped roofs over smooth walls
and rusticated quoins. Between these, the steep slate roof is
pierced by three round *œil-de-bœuf* windows, centered over
dormers whose shouldered, arched frontals endow them with
a slightly Oriental air. Below these the central wall area is
plain, relieved only by a circular window and two rectangular
ones, all carved with keystones to set them off.

At second-floor level the flat roof of the vestibule extends
outward in line with the pavilions, along with the ground-
floor wall in which the main entrance is centered. In this area
the principal decorative features are concentrated, including

The lower terrace and
fountain overlooking the river.

281

OPPOSITE: The drawing room mantelpiece is a Versailles pattern
in white marble.

Another view of the terrace
with the ballroom beneath it.

The garage is as carefully
designed as the mansion.

Pavillons terminating the
grande allée.

the round-arched, carved overdoor with its basket of flowers, fleurs-de-lis in the spandrels, and the tree of life above the keystone, all surmounted by a scroll pediment with its flame finial. The wrought-iron door, too, is a thing of beauty with its brass fittings, its stile bases, with their *cache-pots* of flowers, flanked by life-size bronzes of alert-looking Great Danes by the French sculptor Gardet. Below the upper driveway a shallow flight of steps, adorned with flower-filled urns and hemmed in by thick beds of juniper, leads down to the main roadway through the grounds.

Inside the main entrance is the vestibule with its floor of Etowah pink marble, its carved overdoors, and a cove ceiling supported by arches on carved corbels that at first glance appear to be heraldic shields. At one end is a second wide doorway which admits daylight to illuminate a statuary niche in which a fluted stone pedestal supports a French medieval wood carving of the Virgin and Child. Around this are grouped masses of plants and flowers and a pair of giant South Sea *Tridacna* shells, with an Italian hanging lantern overhead. Against one wall is a pair of carved Régence stools in pink cotton damask, and on the other a French *mille-fleurs* tapestry hangs over an antique oak chest, with a small Aubusson rug on the floor nearby, the whole an exciting promise of things to come.

From this vestibule a pair of six-paned glass doors open into

A colonnade connects
the *pavillon* with the house.

The gallery looking toward
the drawing room.

the stair hall, where not only the floors but the entire winding
stair, including the inside stringer, is of pink marble, the latter
surrounded by a wrought-iron, brass-railed balustrade. In the
hall itself (which is covered with painted canvas in beige with
an alternating green tree and gold fleur-de-lis design) stands
a Louis XV marquetry commode with bronze escutcheons
and mounts and a yellow Siena marble top on which are two
Sèvres vases. Grouped with this is a pair of Régence stools in
leopard skin, with bronze ornaments on knees and aprons and
pieds-de-biche sabots.

On the stair wall the painted design is confined to fleurs-de-
lis, though a pair of floral sprays is added beneath the round
window that marks the center of the stair's sweeping turn.
Into this rounded stairwell hangs a French lantern of gilded
bronze, suspended from an upper ceiling, below which the
decor is completed by an antique Aubusson rug in yellow, gold,
and deep pink.

Even more exciting in the manner in which its decorative
problems have been handled is the gallery. This actually is a

285

OPPOSITE: The entrance hall.

The *sous-sol* powder room.

broad hallway, extending from the drawing room to the dining room, with an opening from the stair hall at its center. This gallery is wholly paneled in quarter-sawed oak, with an off-white ceiling and a parquet floor of an intricate Versailles pattern. Each of the three rectangular doorways is finished with an overdoor panel carved in relief and gilded. The rest of the side wall space, however, is taken up by round-arched panels above a paneled dado except where they coincide with the Roman-arched French windows opening onto a marble terrace, in which case the panel itself is entirely omitted, and in its place draperies of gold silk brocade are hung. Furthermore, the two solid panels between the doorways are fronted by marble-topped consoles supporting arched pier glasses. An extraordinary feature of these mirrors is that each has a three-light bronze appliqué attached to its center in the same manner as on the wooden panels.

Since this room is also a passageway, its center is devoted solely to a long Aubusson carpet, the furniture being arranged along the walls. This consists mostly of Louis XVI pieces—canapés in rose-satin brocade, a tapestry-covered marquise, and *fauteuils* in gold velvet with a delicate floral pattern. Standing in the center window recess is a marble sculpture, *Europa and the Bull* by Paul Manship.

At the drawing room end of the gallery there is a red-enameled Chinese cabinet inlaid with ivory, on each side of the

Inside the entrance foyer.

A reception room in the *sous-sol*.

At the door to the *sous-sol*
reception room.

doorway, containing polychrome figurines. At the other end is a pair of elaborately carved and gilded Florentine vitrines lined with scarlet and white velvet against which a collection of dark-blue patterned Limoges china looks its lovely best.

In the drawing room itself all the wall panels are rectangular, with heavy moldings around the windows. Over these are scroll-ended, brown-painted panels, carved in relief and gilded to form a pleasant contrast to the cresting of the gilded overmantel mirror. The window panels also give the effect of lambrequins, though the gold-patterned draperies are actually enframed by the window moldings.

The mantelpiece is of a Versailles pattern in Poanazetto white marble, garnished with a French clock and a pair of Meissen statuettes. The chenets decorating the hearth are of the Louis XVI period, while the wall decoration is provided by four large tapestries, the one facing the fireplace being a signed Gobelin piece by I. Fourie. Together they form a perfect foil for the colorful Aubusson rug.

The dining room, at the other end of the gallery, is furnished in a manner entirely different from either of the foregoing, having walls of French brocade in a Du Barry pink and six huge, modern mirrors, without frames, which do so much to make the room seem even larger. Though this room is far from small, the horizontal is emphasized with advantage both by the deep, carved cornice, whose crown molding is painted white to match the ceiling, and by the line of the paneled dado. These counterbalance the verticals of the projecting chimney breast and the Louis XVI mantel-mirror combination.

The Louis XVI-style table, with its white top, dominates the room in contrast with the gold damask of the draperies under the carved wood lambrequins and the Louis XVI chairs in a rose-colored damask, both of which look exceedingly well with the wall covering as well as with the Aubusson rug in its gay pink and beige. The total effect, therefore, is one of dignity with overtones of friendly exuberance, satisfying even to the casual visitor.

In this *rez-de-chaussée* section of the house there is only one other room that needs to be discussed briefly in order to illustrate the full range of the quality and interest inherent in the various styles of furnishing and interior design—which are far from monotonous—among these "daylight" apartments, before turning attention to those remarkable rooms below that never see the sun, yet are always welcoming and bright, though representative of styles and fashions that extend back at least 500 years. This third-floor room is the library, which is done in the Gothic style, originating in thirteenth-century France but not coming to fruition until the fifteen hundreds. In any event it is totally different from anything else in the mansion, either aboveground or below, with its pointed arches, linenfold paneling, tracery and finials mostly copied by local artisans from artifacts salvaged from a château near Châlons, plus original locks, keys, and hinges of the same provenance.

The *sous-sol* dining room.

The formal dining room.

The *sous-sol* reception room.

Furnished with a modern sofa as a matter of comfort, this room also contains such things as Henri II (1547–1559) chairs with ram's-head terminals to the arms and a massive carved table of the same period. The coved ceiling is white, and the wide plank floor partially covered by a stunning Oriental rug in red, blue, yellow, and black is punctuated by a bear-skin rug at each end. But it is the carving and blind tracery of the bookcases, the chimney breast and fireplace surround, the door casings, and the carved frieze below the flat central ceiling panel that give the room its monastic character. In addition there are some wonderful French carvings of old-time notabilities and stained-glass panels by Holbein depicting Henry VIII, Jane Seymour, Sir Thomas More, and others of the period.

Added to the richness of the wood, there is plenty of color in the room. Set into the window recesses are gold cotton draperies and window seats with greenish-blue upholstered cushions, while the embrasures are outlined in green, a color repeated in the velvet of the ram's-head chairs. In contrast, the heavy upholstered pieces are covered in crimson, with gold velvet cushions on the sofa. This rich mixture supplies a wonderful feeling of warmth and vitality to the whole room.

Descending from the third floor to the *sous-sol* is something of an adventure, for belowground one depends almost wholly on artificial lighting, though there are several stained-glass windows through which daylight filters from exterior wells. This is augmented by an unobtrusive ceiling fixture and tall bronze *torchères* that flood the white ceilings with light.

Immediately at the bottom of the marble stairs, however, one finds oneself in a tiny hall with walls of huge limestone blocks and a supporting stone beam across the deeply coved ceiling. On the floor is a splendid Oriental rug and pedestals made from medieval wine-press screws supporting wooden statuettes of Saint John. Also, by the entrance to the *sous-sol* lounge is an early seventeenth-century Italian *cassone*, richly carved all over. Above this hangs a gorgeous *mille-fleurs* tapestry from whose floral jungle peer out an astonished-looking lion and a unicorn. On the *cassone* is an antique bronze studio model of a Great Dane dog by Gardet.

Opening the exquisitely carved, arch-topped double doors (the creation of a Yugoslav artist), the even more impressive reception room is revealed, its brick walls inset with diagonal strips of cream-colored stone forming diamond-shaped panels, after the fashion of the Manoir d'Ango of Normandy. Its small fireplace is flanked by a pair of roosters carved in stone, while the mantel serves as a setting for the portraits of the father and grandfather of Joseph Desloge. In one wall recess is a seventeenth-century French chest carved with draped female figures in the twelfth-century manner, its lid the base for a display of twin angelic statuettes in gold, surveying a *cache-pot* of simple blooms. There are two balconies overlooking the lower level, one of ornamental iron, the other a curved stone slab pierced by conventionalized fleurs-de-lis. This and the other *sous-sol* floors are paved with light and dark strips of cork in a parquet pattern of squares to match the diamond-

A musician's balcony
in the *sous-sol*.

A general view of
the *sous-sol* lounge.

Detail of the lounge fireplace.

The Gothic library.

shaped pattern of bricks in the walls. The antique Oriental rug is red and gold.

Down the stair is the *sous-sol* lounge with its massive carved stone fireplace, copied from one at Amboise, which features heavy brackets and a carved scroll pediment behind which the stone hood, adorned with fleurs-de-lis and a wide acanthus-leaf frieze, touches the ceiling. Here, also, is another *pièce de résistance*, a structure formed of hand-hewn oak beams and joists with carved scroll brackets, resting on stone corbels, carved with the heads of winged cherubs. Flanking the fireplace is a pair of Henri II ram's-head chairs with crimson velvet seats, facing on an Oriental hearth rug which has a gold-and-red-striped sofa at each end of a larger one, opposite the fireplace, in crimson velvet. Upholstered seats along two of the walls are covered in the same striped fabric.

Above the white stone dado are four matched windowlike recesses, the two flanking the fireplace paneled in scarlet brocade as a setting for tall, silver trumpet-shaped vases filled with flowers.

Adjoining the lounge is the dining alcove, entered through a broad central arch. Here the two windowlike panels are lined with rich blue-and-gold velvet brocade, and in one of them is a lovely fifteenth-century Florentine tribune showing five charming ladies, exquisitely carved in polychrome, enjoying a knightly joust. The side chairs are painted and gilded Louis

The stunning library
fireplace detail.

XV pieces, both oval and square-back types, the latter in either gold leather or red, the oval backs in tapestry.

The huge, round-ended dining table has a border carved in the egg-and-dart pattern and heavy carved pedestals, the top being normally furnished with a great silver epergne whose mirrored base supports both porcelain water lilies and Portuguese ecclesiastical figurines in polychrome.

At either end of this is a pillar-type candelabrum, each holding six tapers, their light augmented on occasion by small, wall-type lanterns. In one corner, beside the stone stair, is a quadrant-shaped musicians' gallery, set high between dado and ceiling, the curved balcony front pierced by fleur-de-lis-shaped holes centered in the latticelike, diamond-patterned stone ribs. On one side is a polychrome panel, showing Christ disputing with the Elders in the temple; on the other side is a medieval tapestry above a modern piano—the whole thing a tremendous and decorative *coup de maître*, combining the old with the new in a manner it would be hard to parallel anywhere, at any cost, in this day and age.

ACKNOWLEDGMENTS

The authors wish to express their gratitude to the following individuals—owners, architects, curators, officials of historical and preservation organizations—who not only supplied the information herein, together with many of the illustrations, but also checked the descriptive material for accuracy before publication, and in other ways transformed a normally arduous task into an enjoyable experience long to be remembered:

Mr. and Mrs. Jay P. Altmayer; Mrs. Halstead T. Anderson; Mr. August C. Belden; Mrs. Ludlow Bull; Mr. and Mrs. Charles J. Byrne; Mrs. Emilia B. Cresswell; Mr. Benjamin H. Davis; Mrs. Katharine B. Dehler; Mr. Hugh DeSamper; Mr. Joseph Desloge; Mr. Jerome A. Eaton; Mr. and Mrs. Ashby Farrow; Mr. Harry L. Gladding; Mr. Samuel M. Green; Mr. Louis T. Griffith; Mr. George W. Hamer; Mrs. Fred M. Hodges; Mrs. Philip D. Laird; Dr. and Mrs. Charles W. Mayo; Mr. Myrlin McCullar; Mr. Robert A. McGaw; Mrs. E. N. McKinnon; Mr. Joseph N. Minissale; Mr. Joseph P. Monigle; Miss Lucy Morgan; Mr. William P. Osgood; Mr. Robert S. Pflueger; Mrs. Laurence W. Polson; Mrs. William P. Roth; Mr. Floyd E. Shaw III; Mr. and Mrs. Frank G. Strachan; Mr. Harold P. Talbot; Mr. and Mrs. Milton R. Underwood; Mr. David B. Warren.

294

Bayou Bend: F. Wilbur Seiders, 82–84, 86 bottom, 87, 88, 89 bottom, 90–93; Ed Stewart, 85 bottom, 86 top, 89 top.

The Minissale Residence: Lionel Williams, 101, 102, 103 top; Mel Schieltz, 98–100, 103 bottom.

The Hay House: Greater Macon Chamber of Commerce, 104; Ralph Jones, Drinnon, Inc., 105 bottom, 106–110; Max Hunn, 111.

Marble House: Richard Merrill, 112, 115, 117, 119, 120, 121 bottom; Preservation Society of Newport County, 113, 114 top, 116, 118, 121 top.

The Alsop House: Lionel Williams, 122, 123 top, 124–129.

The Read House: Brooks Studio, 130; J. L. Hamar, Jr., 133 top; Sanborn Studio, 133 bottom; Lionel Williams, 134 top, 135, 136, 137 bottom, 139; Mary Eleanor Browning, 134 bottom, 137 top, 138.

The Thomas-Jencks-Gladding House: A. Aubrey Bodine, 140, 141, 142 bottom, 143–149.

Westbury House: Old Westbury Gardens, 150, 151; Denise Otis, 152 bottom, 157 top; Gottscho-Schleisner, Inc., 153, 154, 155 bottom, 156 bottom, 157 bottom; H. Conroy, 155 top, 156 top.

The President's House: Jerome Drown, 158, 159 bottom, 160–165.

Stan Hywet: Bill Samaras, 166–168, 169 bottom, 170–173.

Filoli: Moulin Studios, 174–177, 178 bottom, 179–183.

The Hammond-Harwood House: M. E. Warren, 184, 185 bottom, 187 bottom, 188 bottom, 190 bottom; J. H. Schaeffer and Son, 185 top, 190 top; Francis E. Engle, 186 bottom, 187 top, 188 top, 189.

Oatlands: Allen Studio, 191, 192 top, 193–196, 198 top, 199, 200; Nelson C. McClary, 197, 198 bottom.

The Vanderbilt Mansion: W. Ray Scott, National Park Concessions, Inc., 201, 205, 209 top; National Park Service, 202 top, 206–208, 210, 211; Ralph Anderson, 203, 204; Vanderbilt Mansion, 209 bottom.

Cherokee: Lionel Williams, 212–214, 215 top, 217, 220; Prince Studio, 215 bottom, 216, 218.

Chinqua-Penn Plantation House: Martin's Studio, 221–224, 226, 228–230, 231 bottom; *Greensboro Daily News,* 225, 227, 231 top.

Eleutherian Mills: Hagley Museum, 232, 233, 235–243; Eleutherian Mills Historical Library, 234.

Rosedown: House & Garden, 251, 252; Max Eckert, 244, 245 bottom, 246–250, 253–255.

Palmetto Hall: Parrott Photography, 256, 257, 258 bottom, 259–266.

The Governor's Palace: Colonial Williamsburg.

Vouziers: C. J. Laughlin, 278, 281 bottom, 282, 283; Piaget Studio, 279, 280, 284–293.